Sir Arthur Helps

Some talk about animals and their masters

Sir Arthur Helps

Some talk about animals and their masters

ISBN/EAN: 9783337229405

Printed in Europe, USA, Canada, Australia, Japan

Cover: Foto ©berggeist007 / pixelio.de

More available books at **www.hansebooks.com**

SOME TALK ABOUT

ANIMALS AND THEIR MASTERS

BY THE

AUTHOR OF 'FRIENDS IN COUNCIL'

STRAHAN & CO.
56, LUDGATE HILL, LONDON
1873

[The right of translation is reserved]

TO THE

BARONESS BURDETT COUTTS

WHOSE EFFORTS TO PROMOTE THE HUMANE TREATMENT OF ANIMALS
HAVE BEEN EARNEST AND UNREMITTING

This Work is Dedicated

WITH MUCH REGARD AND RESPECT

BY

THE AUTHOR

Nous debvons la justice aux hommes, et la grace et la benignité aux aultres creatures qui en peuvent estre capables : il y a quelque commerce entre elles et nous, et quelque obligation mutuelle.—MONTAIGNE.

Du führst die Reihn der Lebendigen
Vor mir vorbei, und lehrst mich, meine Brüder
Im stillen Busch, in Luft und Wasser kennen.
            Goethe.

The gentleness of chivalry, properly so called, depends on the recognition of the order and awe of lower and loftier animal life, first clearly taught in the myth of Chiron, and in his bringing up of Jason, Æsculapius, and Achilles, but most perfectly by Homer, in the fable of the horses of Achilles, and the part assigned to them, in relation to the death of his friend, and in prophecy of his own. There is, perhaps, in all the 'Iliad,' nothing more deep in significance—there is nothing in all literature more perfect in human tenderness, and honour for the mystery of inferior life—than the verses that describe the sorrow of the divine horses at the death of Patroclus, and the comfort given them by the greatest of the gods.—RUSKIN.

SOME TALK ABOUT ANIMALS AND THEIR MASTERS.

INTRODUCTION.

THE conversations that follow took place during an Easter vacation. The persons who joined in the conversations were those who have before been known as 'Friends in Council.' They are Sir John Ellesmere, a lawyer of much renown; Sir Arthur Godolphin, a statesman and a learned man; Mr. Cranmer, an official person; Mr. Mauleverer; Mr. Milverton; Mrs. Milverton; Lady Ellesmere; and myself, Mr. Milverton's private secretary. It was sometimes their fancy to take one theme as the subject of their conversation; and this would be kept to as closely as the discursive nature of some of them, notably of Sir John Ellesmere, would allow.

The reason why the particular subject of the treatment of animals was chosen on this occasion is truly related by Mr. Milverton, who, after his escape from drowning, said to me exactly the same thing which he tells to the other friends.

I cannot help thinking that the general question is one of the deepest interest, and that it is one which may be well treated in the way of dialogue. I at first urged Mr. Milverton to write an essay, or treatise, on the subject; upon the whole, I am glad that he did not adopt my advice, but brought out the points which he wished to urge in the course of these conversations with the other 'Friends in Council.'

CHAPTER I.

Mr. Milverton. I want to consult you about something. But, first of all, I must tell you that there has nearly been a vacancy among the 'Friends in Council!' I was upset from a boat in the river the other day.

Sir John Ellesmere. Good gracious, Milverton! How could you be so foolish as to let such a careless person as yourself go out in a little boat—for I have no doubt it was a little one—on this perilous river?

Mr. Cranmer. I must say it was very imprudent.

Mr. Mauleverer. One optimist the less: what a loss to the world!

Ellesmere. But tell us all about it.

Milverton. My godson, Leonard Travers, was going out to the colonies; and the day before he went, he asked me to go out for a row with him. I hate boating: one can't move about in a boat. What Dr. Johnson says of a ship is, to my mind, still more applicable to that greater evil, a boat. But when a young fellow is going away, and one may never see him again, one can't refuse him anything.

Ellesmere. He was not so idiotic as to let you steer, was he?

Milverton. No: but I winked, or coughed, or pointed to some beautiful building on the side of the river, whereupon the wretched thing—I think they call it an outrigger—turned over; and there was I in the water. Luckily it was not far from the shore, and somehow or other I got to land, having been immersed from head to foot. Not one of the least annoyances on such occasions, is the being accompanied by a troop of boys to the first place of refuge.

Ellesmere. That is one of the most curious facts in natural history. In tropical climates an overladen mule falls down upon the sandy plain, never to rise again. Forthwith, in the dim distance, a black speck is seen to arise. It is the vulture which is coming for a feast. There is the same phenomenon to be observed with boys as with vultures. I met with a cab accident the other day. The axle broke, the wheels came in, on both sides of the cab, and we were at once a pitiable wreck. Thereupon, twenty or thirty boys, appearing to rise out of the ground, surrounded us. It is my firm belief that misfortune breeds boys without any superfluous assistance from parents.

Milverton. I must now tell you what were

my thoughts, after my first thankfulness for deliverance from what was really a great peril.

I have written several books in my lifetime.

Ellesmere. Yes!

Milverton. And have discussed many subjects in those books.

Ellesmere. Yes!

Sir Arthur Godolphin. You need not speak in quite so dolorous a tone Ellesmere.

Ellesmere. It is always painful to listen to one's friends' confessions of their past errors and follies.

Milverton. Never mind his nonsense. What I was going to say is, that I have never done justice to a subject which has the deepest interest for me, namely, the treatment of the lower animals by man. I said to myself, I will not go out in a boat again, or take a journey upon the —— railway, before I have put down my thoughts, and really said my say, upon the important question of the treatment of animals.

Mauleverer. But what did you want to consult us about?

Ellesmere. What a farce the consultation of friends is! What a fellow generally wants, and is very angry if he does not get when he consults his friends, is an entire approval on their part of what he is resolved to do.

Milverton. I wanted to consult you as to the best means of putting forward my views upon this question. Shall I try a pamphlet?

Ellesmere. No; people can't abide pamphlets in these days. The pamphlet has vanished into space.

Sir Arthur. More's the pity. Some of the best things that were written in our early days were put in the form of pamphlets. Do think of Sydney Smith's pamphlets, for instance.

Ellesmere. But pamphleteering is a dead and gone thing.

Milverton. I could introduce what I want to say into some report.

Ellesmere. There is nothing so confidential as reports. If I wished to make love to a lady, and to make it most secretly, I should insert my love-letters into some official report, and then get it published in a blue book. Why not talk the matter out? I know that conversations, even ours, are a perplexity to some people—those people who are always anxious for clear, undoubted views, for definite results, for something at once to enlighten and guide them without any trouble on their part. But I boldly say this, that the greatest and most secure portion of the teaching of the world has been done in and by conversation, or, to use a finer word, in and by dialogue.

Here Sir John Ellesmere paused, and there was silence for a minute or two.

Sir Arthur. I never heard this question so boldly stated; but, upon my word, I think Ellesmere is right. Many of the most memorable things in literature, and even in higher teaching than that of literature, have been given forth in dialogue.

Milverton. Dialogue has its drawbacks; but I think with Ellesmere that it has immense advantages. It happens particularly to suit me, because I am always anxious not to overstate, and to be tolerably secure in what I ultimately make up my mind to abide by. Now if I submit any of my thoughts to you, who are men of such varied natures and pursuits, and these thoughts pass muster with you, or do so without a damaging amount of objection, I feel tolerably comfortable about them, and think that they may then be given to a wider circle. But I am not ready now.

Ellesmere. Yes, you are. If we allow you to get 'ready,' as you call it, there will be a treatise. How is it that most pictures are spoiled—especially portraits? By working too much at them. I have often observed that there is a great likeness after the third sitting, which is gradually improved away. It is the effort at completeness which results in that 'padding' that is the ruin of so much good work. Give us your main thoughts, if only the headings that there would be of chapters.

Milverton. First, I should point out the enormous extent of thoughtless and purposeless cruelty to animals. You really can have no adequate idea of this, until you have studied the subject, when you will be able to appreciate the vastness of this area of cruelty. The subject would be divided under several heads: the cruelties inflicted upon beasts of draught and of burden; the cruelties inflicted in the transit of animals used for food; the cruelties inflicted upon pets; the cruelties perpetrated for what is called science; and, generally, the careless and ignorant treatment manifested in the sustenance of animals from whom you have taken all means and opportunities of providing for themselves. It is a formidable catalogue, and I think that the details which I should furnish for each chapter would astonish and shock you beyond measure.

Sir Arthur. Doubtless you have considered, in reference to this subject, the varied treatment of the lower animals, by the different races of mankind.

Milverton. I have; but I think that there is a broader way of looking at this part of the subject than that which has reference alone to difference of race. There is no one phrase which would embrace what I mean; but, speaking generally, the difference of human conduct to animals depends largely upon the differences of culture in men, and still more upon the differences of their familiarity with animals.

I am very glad, Sir Arthur, that you asked me the question which you have just asked; for it brings me naturally to a mode of viewing the subject, which seems to me of the utmost importance, and which I did not see the proper way of introducing. Now let us go into detail. When the familiarity is extreme—when, for instance, the lower animal is constantly the companion of man, and is one of the family, as, for example, the horse with his Arab master—the man begins to understand the lower animal; and understanding of this sort necessarily produces kindness and sympathy.

There is a familiarity of a much lower order, and this does not necessarily produce kindness, unless it is accompanied by some culture.

Then, there is culture of a high kind, such as exists in the higher classes everywhere. That amount of culture would lead to a thoroughly good treatment of animals, if it were but joined with the needful familiarity.

There is always something rather hazy in any axioms of a general kind that one may lay down. A very slight, yet significant, illustration will carry home my meaning to you. **There is a thing called the bearing rein.** It is an **atrocity when applied to** a draught horse. It contradicts **every sound principle** connected with this subject. The coachman, who has **some** familiarity with the **animal, but not the** Arabian familiarity, is uncultured, and **has not the** slightest notion of the real effect of this rein. **The cultivated** master or mistress, **who knows, or might by a few** words **be** taught, the mischief **of this rein,** and **the** discomfort which **it** causes to the animal, is often so unfamiliar with the animal, that **he or** she is quite unobservant of the way in which it is treated, and does not understand its mode of expressing its discomfort. You will notice, on the other hand, that, as a general rule, the educated man who drives **his** own horses, **and** learns to know something about them, slackens this bearing rein, or leaves it off altogether.

Now this comparatively trivial instance is capable, I can **assure you, of the** most general **and** wide application. The one class does **not know,** the other does **not heed.**

That most accomplished **of** modern political economists, **Bastiat,** points out how **a** class **of** work falls into routine, **and** into the sphere **of action of** the least instructed **classes :—**

Un ensemble de travaux qui suppose, à l'origine, des connaissances très-variées, par le seul bénéfice des siècles, tombe, sous le nom de *routine*, dans la sphère d'action des classes les moins instruites ; c'est ce qui est arrivé pour l'agriculture. Des procédés agricoles, qui, dans l'antiquité, méritèrent à ceux qui les ont révélés au monde les honneurs de l'apothéose, sont aujourd'hui l'héritage, et presque le monopole, des hommes les plus grossiers, et à tel point que cette branche si importante de l'industrie humaine est, pour ainsi dire, entièrement soustraite aux classes *bien élevées*.*

This remark, as you see, applies to agriculture. I am going to apply it to the treatment of animals. By the way, I must just note that Bastiat's censure does not apply to England so much as to France; for it cannot be maintained that with us agriculture is the monopoly of the most coarse men. This correction, however, of Bastiat's statement will not detract from the force of my illustration, but will enhance its value. If you had the intelligence of cultured people, joined to the familiarity with animals which the ordinary practical farmer possesses, you would then have an admirable treatment of stock, and that includes a humane treatment. Pretty nearly half the diseases of the domestic animals are the result of a direct violation of the laws of nature upon the part of the owners of the animals. The loss to the nation is immense; and I am convinced that there is no way out of this

---

* Bastiat, 'Harmonies économiques.'

difficulty, but by the union in the owner of culture and a certain familiarity with the animal which he owns.

What I have just said is substantially the answer to your question, Sir Arthur. You ask me to explain the varied treatment of animals by different races. I reply to you—Do not look at the question as a matter of race. There are higher laws which govern it than those resulting from difference of race. I would not, however, pedantically lay down as a dictum that race has nothing whatever to do with the matter. Difference of race may have some influence, but not that dominating influence which the other causes I have intimated possess.

It is always pleasant to indulge in a little personality. Here is Ellesmere. I suppose every one who knows him would admit that he is fonder of the lower animals than of men—at least he finds much less fault with these lower animals. But he is not familiar with them, except with our dog Fairy, and his equestrian knowledge is not quite on a par with his legal knowledge. I might venture to assert that he has not driven a pair of horses since he left college. I observed that his coachman was as absurd as most other coachmen about this detestable bearing rein. When Ellesmere was first made Attorney-General—

Ellesmere. How I do hate personality!

Milverton. —the bearing rein was tightened in honour of the master's rising fortunes. Poor Ellesmere never noticed this. What is the good, you see, of fondness for animals without knowledge or observation of their ways? I hate interference with other people's affairs, but I could not stand this tightening of the bearing rein, and so I attacked Ellesmere himself upon the point. Moreover, I ventured to discuss with him the arrangements of his stable. They were abominable. Of course his horses were always ill. Ventilation was a thing unthought of. I must do Ellesmere the justice to say that he listened to me very patiently, and provided remedies for all the evils I noticed. Now is not this a true bill, Ellesmere?

Ellesmere. Yes: I must own it is; and I wish you could know the trouble I had in persuading my coachman to discard the bearing rein.

Milverton. But, seriously speaking, though my instances may have been trivial, are they not sufficient? I should not like to worry you with all the details which go to prove ill-management, both as regards economy and humanity, of draught horses and of farming stock generally. I re-state my first statement, which is to this effect: that perfect familiarity with the animal will almost supply the place of culture; that imperfect familiarity requires to be joined with culture on the part of the owner; and, finally, that culture **without** the requisite familiarity admits of barbarous things being **done, or rather permitted,** by the owner, from sheer want **of** thought and observation.

Ellesmere. He lays down the **law, doesn't he?** One would think he had been brought up as a farrier, a veterinary surgeon, or a cow-doctor.

Milverton. **I hope** I have not been arrogant; **but** though not a farrier nor a cow-doctor, it has been a part of my business, for many years, to consider the treatment of animals—a somewhat hard fate for one so sensitive as I am as regards their sufferings. But, perhaps, the knowledge I have gained may be made of some use, and so I must not mind the pain that I have endured in gaining it.

Sir Arthur's question diverted me **from the branch of** the subject **I** was next going to consider; which was the transit of animals. The cruelties **of this** transit have increased, by reason of the changed modes of locomotion. Perhaps, however, **it** would be safer to say **that new** forms of suffering have been introduced by this change, to alleviate which the proper remedies have not yet been fully provided. **The** Transit **of** Animals Committee, of which I was a member, made a beginning in the way of improvement as regards this transit; and their recommendations have been, to some extent, adopted by the

Government. But much remains to be done. I will give you the most recent case within my knowledge of the inhuman treatment of animals in transit.

A few weeks ago there arrived at one of our ports a German vessel, with 1,793 sheep alive, and 5 dead. On inquiry, it was found that 646 sheep had been thrown overboard during the voyage. Of course this was put down to stress of weather; but the real truth was, that the poor animals had been most inhumanely crowded together, without any of those provisions against over-crowding which were laid down, as absolutely necessary, by our committee. In fact, the 646 sheep were suffocated. You will observe it was a German vessel, and all that our Government could do was to lay the facts before the German authorities, expressing a hope that when the attention of the German Government should be given to the regulations which the English Government have adopted in this matter, and to the cruel sufferings which are experienced by animals during transit through over-crowding and faulty ventilation, steps might be taken to compel shipowners to adopt contrivances which would lessen the amount of ill-usage to which animals are frequently exposed during their passage from Germany to this country.

Ellesmere. Bismarck would soon set these matters to rights, if he once gave his attention to them.

Cranmer. **But** surely, Milverton, the remedy for these inhumanities, which result in such great losses, will **be provided by the shipper?**

Milverton. There enters the peculiar delusion which besets men like yourself, **Cranmer, who believe wholly** in certain *dicta* **of political economy. You think,** or you talk as if **you thought, that every man** has a plenary **power** of protecting **himself and his own interests, whereas** I maintain that the individual **is often perfectly powerless.** The owner of those sheep, **doubtless,** grieves over his loss—is perhaps half-ruined by **that** loss, and would, as you contend, take care not to run the **risk of any such** loss again. But in practice this is not found to be the case. A man brings his sheep to the place **of** export. **He** cannot afford to keep them long there, waiting for a vessel **that** should be properly prepared to carry animals, if indeed there exists **such a** ship engaged in the trade at that port. He is not a shipowner, and has probably but little influence with shipowners. It would require great skill, energy, and devotion to a purpose foreign from his pursuits, to organise a combination of **sheep dealers,** who might **insist upon** provision being made in cattle-carrying ships for the proper treatment of animals.

I made inquiries, **from experienced persons,** upon this **very point.** I asked **whether this** owner of the suffocated sheep would be likely to be able to provide, by any management on his part, against a recurrence of this fearful and cruel loss. They told me that they thought not—that he would be obliged to use the means of transit provided at his port, whether they were good or whether they were bad.

This is one of the many instances in which, as I contend, the only remedy is to be found in Governmental action.

Sir Arthur. I agree with you in the main, Milverton; but you slightly exaggerate here. In the course of time a remedy would be found. Gradually combination would arise among the sheepowners: competition would come in. New vessels would be built, in which, from the first, when it would be far less expensive, arrangements would be made for the more humane carrying on of the traffic.

Milverton. Well, it may be so: but observe all that you say is hypothetical, and is to happen, if ever it does happen, 'in the course of time,' whereas the German Government can prescribe, as our Government has prescribed, certain regulations which would at once go some way to attain the desired object.

You must not expect the owner of the cattle to take excessive heed of the loss that even he himself sustains by unwise and cruel modes of transit. He is accustomed to that loss: he knew of its extent when he first entered into the business : he looks upon it as a result of the nature of things ; and, finally, what is of most importance, he calculates that his competitors suffer exactly the same amount of loss, and therefore it is of but little matter, commercially speaking. It is only when an exceptional loss occurs to himself, that his attention is aroused to the general loss arising from the ill-managed transit of animals. But for the public it should be a matter of daily concern, and of much interest to them, that this transit should be well-regulated. Comparing the whole business with that most fearful and wicked one, of the transit of slaves, their numerous deaths in 'the middle passage' was of very much less importance to the owner of the slave-carrying vessel, than to the slave-importing colony or country.

You cannot doubt that the public suffers greatly from this imperfect and inhuman mode of transit—in the loss of those animals which die, in the deterioration of those that survive, and in the probable introduction of diseased meat. On those grounds alone I think we have a right to ask for the interference of Government. It is very probable that, at the port of entry in England, the price of meat was slightly increased on the next market-day after that German vessel had arrived, by reason of the loss of these 646 sheep that were expected. Many a poor housewife may have grumbled and fretted at this increased price, and blamed the butcher, little imagining that it was all the fault of some Geheimrath, in distant Berlin, who had never taken the pains to look after a matter that properly belonged to his department.

I may also notice how injurious is the absence of the proper regulations for cattle-transit to the exporting countries. Of course their cattle come to be looked upon unfavourably in the markets of the world; and therefore the improvement of their cattle-carrying ships is a thing which concerns the whole of their agricultural population.

Ellesmere. I hate a fellow who is always chock full of facts. No sooner does one produce a good argument (I really thought Cranmer was going to make a good case), than our fact-full friend whips out some unpleasant fact, which knocks over the whole of the argument. I must say that Milverton has the best of the contest.

Mauleverer. I have not hitherto said a word. I know as well as possible that whatever I should say in the way of opposition or cavil, would be met on Milverton's part by some of these unpleasant facts; and so I shall join in the conversation by coming to his aid. I like what you call the lower animals, and though I think that men are nearly incorrigible, something may be done by educating them a little better, in regard to the humane treatment of animals. I am not a great frequenter of preachers now; but, upon a moderate calculation, I think I have heard, in my time, 1,320 sermons; and I do not recollect that in any one of them I ever heard the slightest allusion made to the conduct of men towards animals. I think that it would not have been a wasteful expenditure of exhortation if, in two per cent. of these sermons, the humane treatment of animals had been the main subject of the discourse.

Ellesmere. Very good, Mauleverer. The great defect of preaching nowadays is, that the sermons appear to be built upon the supposition that the preacher is introducing Christianity for the first time to the notice of his hearers.

Milverton. Returning to the treatment of beasts of draught and burden, I sometimes think that it was a misfortune for the world that the horse was ever subjugated. The horse is the animal that has been the worst treated by man; and his subjugation has not been altogether a gain to mankind. The oppressions which he has aided in were, from the earliest ages, excessive. He it is to whom we owe much of the rapine of those ages called 'the dark ages.' And I have a great notion that he has been the main instrument of the bloodiest warfare. I wish men had to drag their own cannon up-hill; I doubt whether they would not rebel at that. And a commander, obliged to be on foot throughout the campaign, would very soon get tired of war.

To what a height of material civilization a nation might arrive without the horse, was to be seen in Mexico and Peru, when the Spaniards first entered and devastated those regions, where they found thousands of houses well built, and with gardens attached to them. I doubt whether there was a single Mexican so ill-lodged as millions of our poor countrymen are. So you see, when I almost regret the subjugation of the horse, I assume that civilization would not thereby have certainly been retarded.

Ellesmere. I do not object to the horse having been subjugated; but what I regret is, that he does not make a noise. Considering how he is wronged, he is the most quiet and uncomplaining creature in the world. Observe the cab-horse quietly lifting up one of his fore-feet, just showing to the observant by-stander how full it is of pain (you see I do observe animals sometimes); and then think what a row any other animal would make in a similar condition, and how noisily he would remonstrate against the needless brutality of his driver. His conduct and its results form a notable instance of the folly of being silent about our grievances. The busy world pays attention only to those who loudly complain, and accords that attention in exact proportion to the loudness and persistency of the complaint. If there had been a Rochefoucauld, or an Ellesmere, among the horses (for, doubtless, like all other animals, they have a way of communicating with one another), what judicious maxims he might have instilled into them!

There have been a few wise horses in the world. I knew one myself of a sorrel colour. He did not kick, or rear, or pursue any of those fantastic devices for getting rid of his rider; but when he objected to him, he always rubbed him off against a wall or a cart-wheel. No human being, who made himself objectionable to this horse, was ever known to '*remain.*' You do not understand the allusion. A Frenchman, who had taken to riding in England, was asked how he succeeded in this mode of locomotion, so novel to him. He replied—

'When he go easy I am (*j'y suis*); but when he jomp hard, I do not remain.'

Now nobody could 'remain' upon the horse I have been telling you about. But, alas! a wise horse, like a wise man, often keeps all his wisdom to himself; and this wise sorrel (was not the wisest horse that Gulliver met with in his sojourn with the Houyhnhnms a sorrel nag?) did not impart his secret to his brother bays or greys.

Sir Arthur. I say, Milverton, what about pets?

Ellesmere. Yes, let us question and cross-question him, and not allow him to keep exactly to systematic discourse. That is the way in which truth is best arrived at.

Milverton. It goes against the grain with me, to speak against the keeping of pets; and for this especial reason, that the young people who keep pets are generally, in after life, those who are the best friends to animals. But, if I must answer the question truthfully, I do think that there is a great deal of cruelty in keeping pets—not so much directly as indirectly. There are the cruel devices by which pets are caught and tamed. Moreover, we make pets of creatures which were never meant to be made pets of. I allude particularly to the feathered creation. A miserable creature, to my mind, is a caged bird. I do not know that I ever saw a countenance more expressive of dignified misery, and of its owner having known better days, than that of an eagle which I once saw in a cage about five feet square. Of course what I have just said does not in the least apply to those creatures, such as cats and dogs, which really appear to like the society of men.

Ellesmere. I am always afraid lest dogs should come to learn our language. If they ever do, they will cut us entirely. Everything seems clever and uncommon-place in a language of which you know but little; and that is why we appear such clever and interesting fellows to dogs. If they knew our language well, would any dog sit out a public dinner? Would any dog remain in the nursery, listening to the foolish talk of nurses and mothers? I am not quite sure whether our Fairy here would stay so resolutely with us, if she understood all we said.

Sir Arthur. There is a fact which militates against your theory, Ellesmere, and that is, that a colley dog understands his master better than other dogs understand their masters, and yet he is true to him, and does not cut him.

Ellesmere. No! it makes for me. The shepherd uses certain signs, and they are sensible signs. They indicate certain judicious things to be done. The dog approves of the proposed transactions, and willingly takes his part in them. He gives his master credit for judicious talk at home, which the dog does not understand, but supposes to be equally clever with that which takes place between himself and his master on the hill-side.

Mauleverer. Going back for a moment to the pets of which Milverton disapproves, I hope that he includes gold-fish. When I see those wretched creatures moving round and round about in a glass bowl, I don't know how it is, but I always think of the lives of official and ministerial people, doing their routine work in a very confined space, under very unpleasant and continuous observation, never suffered to retire into private life amongst comfortable weeds and stones and mud, but always having the eyes of the public and the press upon them.

Ellesmere. That is a very sound simile of Mauleverer's, and it seems to me that I ought to have seen the similitude before. I will treat you to another simile of equal exactness. Whenever I see a favourite cat, with its so-called master or mistress, I always feel that the cat considers the master or mistress as a hired companion. The cat feels that it has somebody to open the door for it, to find out the sunniest window-sill for it, and in fact to perform a thousand little offices belonging to the duties of hired companionship, in return for which the cat purrs out some wages, and is content always to be in a graceful attitude, as an additional payment to the hired human companion.

Sir Arthur. Don't calumniate cats, Ellesmere; I once had a cat which—

Ellesmere. Forgive me for interrupting, but I must tell you something which I may forget to tell you if I do not say it now. The word 'calumniate' puts me in mind of it. It relates to calumny, or rather, perhaps, to scandal. It will be worth the whole of the rest of our conversation to-day.

Some girls were asked by one of our inspectors of schools, at a school examination, whether they knew what was the meaning of the word scandal. One little girl stepped vigorously forward, and, throwing her hand up in that semaphore fashion by which children indicate the possession of knowledge, attracted the notice of the inspector. He desired her to answer the question; upon which she uttered these memorable words: '*Nobody does nothing, and everybody goes on telling of it everywhere.*'

I once read an essay of Milverton's about calumny, which has not been published, I believe; and it was divided into sections and sub-sections, and was meant to be very exhaustive. There was nothing in it, however, equal to this child's saying, which in fact reminds one of Bacon, Swift, and Macchiavelli all compounded together. Listen to it again. '*Nobody does nothing*' (regard the force of that double negative), '*and everybody goes on*' (note the continuity of slander) '*telling of it everywhere.*' (No reticence, you see, as regards time or place.) I am sure that some member of that child's family, father, or mother, or sister, or brother, had been subject to village scandal, and the child had thought over the matter deeply. I have good authority for the story. It was told me just before I came here by Sir George ——.

Mauleverer. Upon my word it is admirable. That child and I should agree in our views of human life.

Ellesmere. But what about your cat, Sir Arthur?

Sir Arthur. Everything will sound so tame, Ellesmere, after your story. It was merely that I had a cat that would walk out with me like a dog, and would sit for hours on my study-table, watching me at work.

Ellesmere. Yes; it was one of those cases, not uncommon, in which the master or the mistress becomes very much attached to a loveable and agreeable hired companion. I never said that cats were devoid of affection, only that they thoroughly understand their superiority to the human beings whom they take into their employment.

No one, without experience of the difficulty, can imagine how difficult it is to follow and report conversation; and how it strays from one topic to another, in the most eccentric fashion. After Sir John Ellesmere had made the last remark, all the 'Friends in Council' seemed to be devoured by a desire to tell remarkable anecdotes, quite within their own observation, of wonderful cats, dogs, and horses, and even of birds that they had known. Two or three of them talked at the same time, chiefly addressing their immediate neighbours; and I failed to get at any connected thread of conversation. I remember that both Dickens and Lord Lytton were spoken of, as having had great appreciation of the intelligence of birds. Then the doings, which were almost incredible to me, of a certain dog were spoken of. This dog had belonged to a young man, known to most of the company present, who had, somehow or other, gone very wrong lately, and had given his parents a great deal of trouble. And then the whole conversation was about him. Sir John Ellesmere took his part; and, speaking generally of families, said that it was wonderful to him to see how well they agreed together, considering the immense difficulties of the position. My chief, who always delights to hear anyone defended, and excuses made for anybody, followed on the same side. I think I can report the conversation from this stage accurately:—

Ellesmere. Here are people shut up together in the same house, having, probably, very different tastes and very different ideas on all matters of human interest, and being so familiar with one another that the forms of politeness have been somewhat broken down; and you expect all these good people to get on well together, merely because there is a close tie of relationship between them. Why! I do not find that a man and his wife always get on well together; and I am told that here there is no duality, but absolute unity of persons.

Milverton. There is one thing greatly in favour of fathers and mothers. They at first seem to be hardly treated by the laws of the universe. They have almost always to restrain, and control, and give unwelcome advice; and, in short, must, to a certain extent, make themselves disagreeable, by the pressure on their own minds from their own experience. There is foolish talk in this day, as there has been in all time, about the singular rebelliousness of young people. I have been hunting through Pliny's Letters to find his views about animals. I did not come to anything that Ellesmere would call a 'good find;' but I came upon a passage relating to the conduct of young people in his day. He is writing a letter about the great grief which he has felt upon the death of a young man, named Junius Avitus— who, according to Pliny, was a perfect young man. But then Pliny goes on to show how different Avitus was from other young men. Please give me the book, Johnson. Here is the passage:—

Rarum hoc adolescentibus nostris. Nam quotusquisque vel ætati alterius, vel auctoritati, ut minor, cedit? Statim sapiunt, statim sciunt omnia: neminem verentur, imitantur neminem, atque ipsi sibi exempla sunt. Sed non **Avitus.**

Of course this juvenile presumption is hard for parents

and seniors to bear; and we must admit that their reaction against it does tend to diminish love on the part of the children. But there is, if I may so say, an aftermath of good harvest for the parents. As the sons and daughters go on advancing in life, they are nearly sure to recognize more and more the worth and affection of their parents. Anybody who lives some time in the world, and has opportunity of observing the development of families, must be struck with this, if he is an observant man at all. For instance; you hear sons, who, as you know, gave fathers much trouble at certain periods of their lives, now speak of those fathers with the tenderest affection. Indeed, I might almost venture to say that every day that passes over a child's head, whether son or daughter, makes him or her more appreciative of the love of parents. The greatest 'scamps,' as you call them, ultimately feel this—more, perhaps, even than the good children; and I doubt not that this very boy we have been talking of, who is at present a great trouble it must be owned, will hereafter speak of his father, to whom he has given all this trouble, in the tenderest of terms.

Lady Ellesmere. It makes such a difference, having children of one's own. I do not think it is any particular instance of one's former conduct that is wont to occur to one's mind. The whole tone of mind becomes altered as regards the parental relationship.

Cranmer. Oh, how true and how sad, Milverton, is what you have just said! I was, what I believe is called, a good son. I never thwarted my father in any serious matter; but I once behaved very ill to him; and the recollection of this has come back to me a hundred times. Though you may not think it, from looking at me now, at the age of seventeen I was a most delicate youth, and there were doubts whether I was not consumptive. My mother died early, and my father played, as far as a man could, the mother's part to me. One of the things he 'led me a life about,' as I then used to express it, was about warm clothing, and I recollect on one particular occasion we had a quarrel (I almost think it was the only one) about a certain garment he wished me to wear, and which I would not. You know how youths do not like to differ from other youths, and are ashamed of being 'coddled,' as they call it. The result was that I was very rude, and resolutely disobedient. As I said before, this little incident (no; I cannot call it 'little') has come back a hundred times to my mind; and, indeed, I seldom look at my father's picture without thinking how ill I behaved to the good man on that occasion. If the opportunity would but come over again, what amount of clothing would I not put on to please him who has gone, and who probably cared for me more than any other human being has since done!

Mr. Cranmer, whom we always look upon as rather a hard man, at any rate as a very dry sort of man, was visibly affected when he told us this little anecdote. Some of us instantly made efforts to turn the conversation into other channels. I believe the weather, that never-failing subject, came in for its share of talk. Then the conversation languished, the ladies made a move, and the main subject of the discourse was postponed to the afternoon.

CHAPTER II.

It was not to be expected that our friends, most of whom are connected in some way with political life, should not at times talk politics. In general, I have avoided giving any of these discussions, which must of necessity be of an ephemeral character. But, this morning political talk led to a discussion which I thought worth recording.

Some public documents had been very severely criticised by some of 'the Friends,' especially by Sir John Ellesmere; when Mr. Milverton, taking up the defence, said :—

Milverton. You all talk as if it were the easiest thing possible to write clearly, and to say what one really means to say; whereas, it is one of the most difficult things in the world. Now let me state the difficulties. In the first place, to write clearly, you must have clear ideas. In the second place, they must be consistent ideas. In the third place, there must be a logical sequence, and the several parts of the arguments must not overlap one another. Then you must know how to construct a sentence, which, to quote a favourite expression of Goethe's, 'is not everybody's affair.' Then there comes the difficulty of the choice of words, greatly added to by the fact that the words you choose must have the same signification in the minds of the person or persons addressed as these words have with you.

With regard to the choice of words, and its telling effect, I should like to give you an example. Please, Ellesmere, hand me down Barrow's Sermons. They are close to you there. Now, listen to this extraordinary passage, which is extraordinary, simply because every word in it is well chosen. I defy you to find a fault in it, as regards the choice of words. And note how well it might be translated into any other language:—

We have but a very narrow strait of time to pass over, but we shall land on the firm and vast continent of eternity; when we shall be free from all the troublesome agitations, from all the perilous storms, from all the nauseous qualms of this navigation; death (which may be very near, which cannot be far off) is a sure haven from all the tempests of life, a safe refuge from all the persecution of the world, an infallible medicine of all the diseases of our mind, and of our state: it will enlarge us from all restraints, it will discharge all our debts, it will ease us from all our toils, it will stifle all our cares, it will veil all our disgraces, it will still all our complaints, and bury all our disquiets; it will wipe all tears from our eyes, and banish all sorrow from our hearts; it perfectly will level all conditions, setting the high and low, the rich and poor, the wise and ignorant, all together upon even ground; smothering all the pomp and glories, swallowing all the wealth and treasures of the world.*

Sir Arthur. It is, indeed, a transcendent passage.

Milverton. In addition to all the other requisites I have named for good writing, there should, throughout, be a divine accuracy.

Ellesmere. I have known many queer marriages among my human friends and acquaintances, many as queer between substantives and adjectives, especially where substantives have been ill-mated with adjectives; but I have seldom known a noble adjective to be wedded to so commonplace a mate as when 'divine' is applied to 'accuracy.'

Milverton. I shall not withdraw the adjective. A very large part of the evils of the world can be traced up to inaccuracy. It is the chief source of misunderstanding, misrepresentation, and of most of those direful words which begin with that unhappy prefix, 'mis.'

Cranmer. I never heard the difficulties of writing so fearfully set forth.

---

* Barrow's Sermons.

Milverton. I must proceed further. I must add, that most public documents are not the outcome of one man's mind, but are the result of much conjoint writing and intermeddling. I will also say, that ours is a most difficult language wherein to write accurately. Again (I do not know when I shall have done with my additions) : you must consider that the writing of these State papers is a very different thing from ordinary literary work.

Sir Arthur. I understand what you mean. I have been reading a great deal of Schiller lately; and I have found only one or two passages which I cannot understand.

Milverton. Wherever there is passionate or emotional writing, the passion or emotion explains so much, and you have, comparatively, little difficulty.

Sir Arthur. That is just what I was going to say. When I take up a German newspaper I am considerably puzzled.

Ellesmere. Yes; and, perhaps, it would be worse if you took up a German State paper.

Milverton. The great difficulty is to write well about matters connected with real life where precision of terms is absolutely necessary.

Please do not think it pedantic on my part, if I venture to lay down what should be the conclusions derived from our talk on this subject. They are three.

1st. We should cultivate, with care, the art of expressing what we may have to say in writing. And this should be done early. The writing of themes is, in my mind, one of the most absurd employments in the world. You ask boys to evolve ideas from their own consciousness, upon subjects of which they know nothing. You might teach the power of expression in quite another way; and in a secure way. Relate facts to them, and make them restate those facts. Give them a portion of history to read, and direct them to make a summary of it. In short, exercise their powers of expression, if you like, to any extent; but do not ask them for ideas.

2nd. Now I come to the work of the grown-up man. Whenever there is work to be done which requires conjoint counsel, and is to be the result of much clashing of opinion, let one man be entrusted with the drawing-up of the ultimate document, which is to be put forward to the world, and to be commented on by it.

Partnership is a good thing in its way, and is available for many affairs, but not for writing.

3rd. Make much of the man who, whether by natural gifts, or, as it is more probable, by careful study, has attained this rare faculty of expressing precisely what he means to say, or what others wish him to say. That man is worth a great deal of money, whether he is to be found in a Government department or a merchant's office. He it is who should write the letters, though he may not be a man capable of forming the most judicious opinions on the questions at issue.

Ellesmere. In order that he may not be overworked, and that all the fine qualities of this admirable man may not be blunted by over-use, let us diminish the frequency of the arrival of the post. I have just found the most lovely passage in an old novel—not so very old, though—published thirty-five or forty years ago.

Cranmer. I thought that when last we separated, it was agreed that at our next meeting we were to continue to discuss the subject of the treatment of animals.

Ellesmere. Oh, yes; but what is the good of having a set subject for discussion if one may not have the pleasure of breaking away from it sometimes? Besides, whatever irrelevancy I may commit, Milverton is sure to be able to show that it bears closely upon the subject in hand.

Sir Arthur. Give us the 'lovely passage' at once.

Ellesmere. It was something of this sort: 'She looked wistfully at the mantelpiece, and sighed when she saw that there was no letter for her, for she knew that the half-weekly post had come in.' I laid down the book, and sighed too, thinking of the happy days when, at some favoured places in England, there was only a half-weekly post.

Mauleverer. Some foolish people are always fancying that they should like to have lived in some age previous to their own, or wishing that they had not been born just yet, and that their time was to come in some subsequent age.

Lady Ellesmere. I suppose I am one of those 'foolish persons.' I do not agree with my husband in his horror at receiving letters, which, as you know, is a grievance he is never tired of mourning over; but I often fancy how delightful life must have been in quieter times, when everybody was not in such a hurry; when short intervals of distance produced great changes of scene; when small things which happened to one were events, and were well talked over and thought over. Now, nothing makes much impression on our—

Ellesmere. Say, kaleidoscopic minds. You just give a shake of the tube, and there is another set of patterns. I don't object to this.

Lady Ellesmere. But surely, John, you would like to have lived in a time when the world was less built over. I often subtract, in my mind's eye, all the new houses at some beautiful spot, and see what it must have been in former days, and then imagine the douce, pleasant life which the few dwellers in that beautiful place must have led.

Mauleverer. An utter delusion. Remember how

strong must have been the hatreds and the dislikes when people lived together in very small communities, and when there was next to no movement from the localities in which they dwelt.

Sir Arthur. As for Ellesmere, his life might have been a *douce* one (to adopt Lady Ellesmere's word), but it would have been a very short one in most of the previous ages of the world. There is no man of my acquaintance who would have been more certain to have been burnt for heresy, or hanged for treason, than Ellesmere. Is it possible to conceive that he would have been able to restrain himself from taking objections to the dominant views of religion and politics, whatever they might have been? and taking objections would have been torture or death, most probably the latter. This thought reconciles me to a post coming in six times in the day, and to the way in which (for I agree with Lady Ellesmere) most of the beautiful spots of the earth have been deformed by modern houses.

Milverton. These preliminary remarks naturally lead up to my subject.

Ellesmere. I told you so. Nothing can keep his subject down.

Milverton. I spoke the other day of a new set of sufferings endured by animals in consequence of new modes of locomotion being invented; but, on the other hand, it must be admitted that a whole series of atrocities in reference to animals has entirely passed away. I allude to the cruel way in which they were tortured and slain for medicinal remedies. You can hardly take up any old work upon medicinal recipes without coming immediately upon some ridiculous mode of cure, in which some inhumanity is to be practised upon an animal to satisfy the superstitious notions of the age. We have in the house a Welsh work, named 'Meddygon Myddfai.' It is full of these horrors.

Sir Arthur. Yes; I have heard of the work. It is supposed to date from the time of the Druids.

Milverton. True; but do you doubt that it was believed in until quite modern times? and even now there are country districts where these atrocious remedies are entirely believed in as articles of faith.

Mauleverer. I am sorry to be obliged to check any joyfulness at some particular barbarity having dropped out of fashion; for, my good friends, I must remind you of the fact that another has entered. What was done by superstition is now done by science.

Sir Arthur. I deny that the cruelties inflicted by science upon animals, are equal in number and extent to those that were inflicted by superstition;

and then, look at the purpose—recollect that in some cases it is to master the diseases of animals that animals are subjected to scientific investigation.

Milverton. Scientific investigation! It is very unlike you, Sir Arthur, to use fine words for the barbarities that go on under the pretentious name of scientific investigation.

This was one of the branches of the subject that I was most anxious to discuss with you. I do not wish to carry my arguments to any extreme; but I declare that I believe a vast amount of needless cruelty is inflicted upon animals under the pretext of scientific investigation. I vow that I think it is a crime to make experiments upon animals for the sake of illustrating some scientific fact that has already been well ascertained. You might as well say that it is desirable to put wretched dogs into the *Grotto del Cane* for the purpose of proving that the air in that grotto is mephitic.

Lady Ellesmere. Surely everybody must agree with Leonard in that proposition.

Mrs. Milverton. Certainly.

Ellesmere. Well, I do not deny it; I think he is right.

Mauleverer. I go further: I don't believe that a single valuable fact has been discovered by any of the tortures which have been inflicted upon animals.

Sir Arthur. I am not prepared to go that length with you. I have a perfect horror of vivisection; the very word makes my flesh creep. But we shall not carry our point (for I take it we are all agreed upon the point) by suffering any exaggeration to enter into our statements.

Ellesmere. What do you propose, Milverton? I mean, what do you propose by way of remedy for this evil?

Milverton. I have very little to propose in the way of direct remedy. There have been horrors in the way of vivisection—especially those perpetrated in France and Germany—against which I think direct legislation might be claimed. But it is very little that direct legislation can do in this matter. We can only rely upon the force of enlightened public opinion. I think women could do a great deal in this matter, as indeed they can in most social affairs.

Lady Ellesmere. It would be quite enough reason for refusing to marry any man, if one knew that he practised any needless cruelties upon animals, whether called by any scientific name or not. My husband may ill-treat me, as you see he does, generally pointing out the foolishness of any remark that I may make; but if he ill-treated animals, I do not think I could endure him.

Mauleverer. After all, then, we are driven to the

surgeon's wives, or sweethearts, for some remedy in this affair.

Milverton. Not altogether. If public opinion were strong in the direction in which we wish it to prevail, no government, no public body could have these cruel and wicked experiments carried on under its sanction. I have looked into the subject carefully, and I have come to the conclusion that the action of this opinion upon public bodies would stop many of the horrors we now complain of. I cannot say any more about this branch of the subject. It is so repulsive.

I now descend into a very commonplace matter relating to beasts of burden. I think that a great deal might be done to alleviate the sufferings of animals, by reasonable and judicious supervision and inspection. Of course I know what will be said directly, in opposition to this proposal, that it is contrary to the laws of political economy. But, previously to going into detail, I want to ask you all a great question which presses upon my mind. It is this: Has not every living creature its rights? I suppose that this proposition may seem somewhat fanciful when applied to animals; but I distinctly hold that every living creature has its rights, and that justice, in the highest form, may be applied to it. I say that a lame horse has a right to claim that it shall not be worked; and just as I would protect one man from being ill-treated by another, so, to use the principle in its widest form, I would protect any one animal from being ill-treated by any other.

Cranmer. I know that I am always in Milverton's thoughts whenever he makes a dead set against political economy and economists. I do not see much use in his dogma that—

Sir Arthur. I do.

Cranmer. —every animal has its rights; but I have no objection to admit it. What then? What practical result do you aim at, Milverton?

Milverton. Well, I say that if only several practical men cared for this subject—namely, the good treatment of animals—as much as we who are sitting in this room do, an effective system of inspection and supervision might be devised for draught-horses employed in any great town—say, for instance, in London.

The mischief in this world is, that statesmen, and men of business, have for many years been greatly employed in arranging where power should be placed, and not how it should be used; and so poor men and poor animals have often had a sair time of it.

I maintain that, with the assistance of my friend Cranmer, I could give the heads of a Bill for the inspection of draught animals in this metropolis, which would prevent an immense deal of the cruelty now exercised upon them. Once direct the attention of men to this subject (and man is a most ingenious animal), you would be surprised to see what good results might be effected.

But now I am going to propose quite a minor matter—a thing at which I dare say you will laugh, but which, I believe, will have a great effect in relieving part of the misery suffered by draught-horses in our metropolis—I allude to cab-horses.

Ellesmere. Stop, stop: I cannot allow this discussion to go on at such a pace. I must go back to the legal part of the question. Every animal has its rights, according to Milverton. Why stop there? Every reptile then: every insect? Do you admit that, you Brahminical personage?

Milverton. Certainly. You may make me ridiculous, or, at least, try to do so; but you shall not make me inconsistent. Look there: you see, at this moment, in front of the open window, a number of flying creatures.

Ellesmere. Why not say flies at once?

Milverton. Because I wanted to state the matter in the most abstract fashion. You see, I say, a number of flying creatures, whirling about in a mazy dance, and, as far as we can judge, enjoying themselves very much, and doing us no harm. They are not even touching any of that 'property' which the lawyers love so well. If you were to kill any of them at this moment, I think it would not merely be a cruelty, but an invasion of right—an illegal transaction.

Sir Arthur. I think Milverton is justified in this assertion. You have no right to attack those creatures. Have you ever observed, by the way, how fond children are of that word? 'You have no *right* to do it.' 'He had no *right* to hit me,' and so on.

Ellesmere. 'Hail, horrors, hail!' Do you see that cloud, not of insects, but of morning visitors coming up the avenue?—and they have seen us too. Have we a right, we men, to slip off, and leave the ladies to receive them? Right or not, I vanish.

So Sir John rushed off, while the rest of us, having some sense of politeness, stayed; and so the conversation upon the animal question was for a time broken off.

CHAPTER III.

THE conversation this afternoon commenced abruptly as follows :—

Mauleverer. We have been talking a great deal lately about the animal creation. I have been thinking how much misery there is among the lower animals arising from fear.

Ellesmere. Nonsense! You are not going to persuade me of that.

Mauleverer. It is true though. I am convinced that they suffer greatly from fear, or rather apprehension; and that is what we human beings suffer most from.

Sir Arthur. We know very little about their joys, their sorrows, or their sufferings. I am inclined to think that there is a large balance of happiness in their favour. I should have taken just an opposite view to Mauleverer's. I admit that they suffer from fear, but very little from apprehension. What men suffer most from is not fear, but care.

> Heard not by the outward ear,
> In the heart I am a Fear,
> And from me is no escape.
> Every hour I change my shape,
> Roam the highway, ride the billow,
> Hover round the anxious pillow.
> Ever found, and never sought,
> Flattered, cursed. Oh! know you not
> Care? Know you not *anxiety*?*

Ellesmere. I suppose everybody has thought at some time or other what creature he should like to be, if the Pythagorean system were true, and we were to reappear—at least, some of us—in some lower form of Being. I wonder what Cranmer would like to be. I don't wish to suggest anything for any other gentleman; but it occurs to me that Cranmer would rather shine as a tortoise.

Sir Arthur. I have long ago made up my mind upon the point. I would be a bird; and, I think, of all birds, the swallow—a travelled creature, who, like Ulysses, had seen many races of men and many cities.

Ellesmere. Milverton would be a bird too—the meditative stork. I knew a stork just like Milverton. I used to watch it from the windows of my inn at a little place—I forget the name—on a river that runs into

---

* Second part of Faust. Anster's translation.

the Rhine. It would remain for hours perched on a rock, standing on one leg and pretending to look for fish; but, in reality, thinking of the queer ways of men, and inventing aphorisms.

Lady Ellesmere. You have not asked us what we should like to be.

Ellesmere. Oh! you will have gone through the worst form of animal life—at any rate, that which is most noxious to man. But if there is any other form to be gone through by you, you will, of course, be butterflies; and how you will look at each other's fine dresses, and say of the empress butterfly, 'I wonder how her husband can afford to let her dress in that expensive way—in all the colours of the rainbow, too!'

Milverton. That makes me think of an anecdote of Thackeray. He was going down the Strand with a friend of mine, and they stopped to look at an oyster-shop. There was a tub of oysters at a shilling a dozen (those were halcyon days for oyster-eaters): there was also a tub of oysters at tenpence the dozen. 'How these must hate those!' exclaimed Thackeray, pointing first to the tenpenny, and then to the shilling oysters. There you have a most characteristic speech of that great satirist.

Ellesmere. I have been thinking over the question, and I have made up my mind what I would be. I think I once told you before. I would certainly be a fish.

> The fish he leads a merry life,
> He drinks when he likes, and he has no wife.

That is my own poetry; at least, I believe so.

Sir Arthur. I think Ellesmere is wrong in his natural history.

Ellesmere. Well, the fish has no wife like the lion, the tiger, or the fox. I should not like to be a lion, and have to come home, after a hard day's hunting, to the lioness and my cubs, without any prey in my paws. My mane would be pulled in a manner that would not be at all caressing. No. I am quite resolved, that if I am to have any choice, I will undoubtedly be a fish. Wait though! I pause to hear what Mauleverer will say; because I will not enter willingly into any form of animal life, if he is to take the same form. He would make all the other fishes so melancholy, that they would turn up their sides, and show the whites of their eyes, and not endeavour to catch any more flies, for he would prove to them that about once in a million times it would be an artificial, and not a real fly. Let us enjoy life in the best way we can; whether we are birds, beasts, men, or fishes, and eschew all those people who delight in melancholy talk, write melancholy novels, that have bad endings, or play melancholy music which I cannot abide. Dismal people are the only people to be sedulously avoided, unless, of course, as in Mauleverer's case, they have transcendent notions of cookery.

Milverton. The whole of your conversation reminds me—

Ellesmere. The three most fatal words in the English language are, 'That reminds me.' You feel certain that a long story, or a terrible disquisition is coming upon you.

Cranmer. I must interrupt. I am not going to be a tortoise, though I admit it is one of the most judicious creatures in creation. I should like to be a beaver. If there is any truth in the theories that Mauleverer is always dinning into our ears (forgive the unparliamentary expression, Mauleverer), about the misery of all creatures endowed with life, that misery may best be met with and conquered by continuous work. Now, I take it the beaver is the most laborious creature of whose habits we know anything. He is never satisfied with his work. I wonder that he is not a chosen pet of mankind, for people who have watched him tell me that he is a very good-natured animal, and a most amusing one. After building a dam, which he will do in a room, for he always supposes that he is to be victimised by a sudden influx of water, he will contemplate the building in the most knowing manner, strengthen the weaker parts, or pull it all down together and reconstruct from the foundation—in fact, he is a ceaseless worker, and a most severe critic of his own work. This must recommend him to your favourable notice, Milverton.

Ellesmere. I won't be a beaver if I can help it One of the errors of this age is, a deification of work for the mere sake of working. I hate fussiness.

Lady Ellesmere. The most fussy man alive!

Ellesmere. The really good man, and the man of beautiful nature, like the good fish, can be idle and innocent too. Show me the man who employs his leisure well, and I will tell you who will go to heaven.

Milverton. Well, now I suppose that you have all said your preliminary say. I shall take full advantage, hereafter, in commenting on what you have said every word of which bears upon my subject. But, to begin at the beginning, Mauleverer spoke of the fear suffered by all animals. By the way, adopting the plan used in Acts of Parliament, of defining certain words to have extensive meanings, when I use the word 'animals' I mean all living creatures except men and women. Now, touching this fear, I maintain that animals are more fearless than man. There are several domestic animals of my acquaintance, which, having learnt the thorough friendliness of the men, women, and children with whom they live, are most remarkably fearless. Seize hold of them suddenly; threaten them as much as you like; they have that perfect confidence in your good intentions that they will bear the threatening gestures with an equanimity and absence of nervousness which are unknown in man. I mention this fact with a view to show how much we might increase the happiness of these animals with which we live, if we were *uniformly* kind to them. I refer you for proof of this to the happy families of animals that lived with Waterton.

Ellesmere. All animals I have known intimately have had a great appreciation of fun; and that is why I like the animal creation so much. If I were to pretend to throw Fairy into the water, a proceeding which she knows that I know she dislikes, she would perfectly understand that this was a mere demonstration, similar to that of an independent member asking a question of a minister in the House, the whole affair having been arranged an hour or two before at the minister's official residence in Downing Street, and Fairy would thoroughly enter into the joke.

I can hardly tell you how much I see in this. It impresses me more than hundreds of those stories showing the sagacity of animals which are current in the world.

Milverton has been wonderfully merciful to us, in

not giving us hosts of these stories. There was one, however, he used to tell (for we have had much of this animal talk before, though not in the same company) which became a favourite with me. It was from some old writer—Barlæus I think was his name—that the story came. There was a fox which had been the death of many a pack of hounds. He managed it in this way. He got them to follow him in full cry towards a ravine: over it he went at full speed: the dogs followed him, and were dashed to pieces. But the fox by no means descended to the bottom of the ravine. He contrived to jump into a bit of brushwood that hung a little way down the precipice. Thence he had some sort of pathway which he could climb up and regain his lair. Doubtless the creature had found out this device by accident. The first time he had probably gone over the precipice unwittingly; had been caught by the brushwood; and had detected this to be the mode of getting rid, in the most complete and summary fashion, of tiresome packs of brutes that were wont to worry him.

What a difference there is between this trick and an animal's showing that appreciation of character and conduct which is to be seen in Fairy's understanding that any threats of mine are mere *bruta fulmina*, and, in short, a bad style of joke.

Sir Arthur. I can confirm what Ellesmere says,

and can give another remarkable instance. I know a dog who, without flinching (by the way I object to a neuter relative pronoun being applied to an animal, and I advisedly say 'who' instead of 'which'), I say I know a dog who will not flinch, or even wink, when his master, or his master's friends, aim a violent blow at him which will come close to his head; but which of course is never meant to hit him, and never does hit him. Still his confidence is something amazing, and he evidently enjoys the transaction exceedingly. It is a joke in which you can see he thoroughly partakes. His master is one who is exceedingly tender towards animals, and the dog knows that he would not hurt him for the world. And thus the animal appreciates that this is a great amusement to the company. Now observe, there would be comparatively little in this transaction, if his master were a showman. We know by what devices, connected with reward and punishment, tricks of this kind may be taught; but no reward follows this exhibition. The dog is perhaps sitting up at the tea-table with the children, when a violent blow of the kind I have described is aimed at him; but he at once enters into the fun of the thing. I really do not think that Ellesmere presses his instance too far when he makes it significant of an appreciation on the animal's part of the character and habitual conduct of his master.

Mrs. Milverton. All that you have just said bears out what Leonard began by maintaining, that the animals with whom we are obliged to live might be made infinitely more comfortable by the removal of fear, and that this is an end which can be easily attained.

Milverton. We begin to teach by blows, which are things very difficult to understand; and then we wonder that we have no hold upon the regard of the animal, and, in fact, that we cannot manage it. Now, many animals, I should say most animals, have a Macaulay-like memory, and certainly never forget early ill-treatment.

What I desire most in our conduct to animals, is some little use of the imaginative faculty. An imaginative person cannot well be cruel.

Ellesmere. Oh! Oh! Your friend, Cortes, for instance—a poet, a scholar, undoubtedly a man of powerful imagination, yet how he treats the natives! just as brutal men among us treat animals.

Milverton. You are wrong, though perhaps my dictum may require some modification. The imagination of Cortes made him fully aware of the sufferings he was inflicting. That same power of imagination led him to believe that he was doing great things for civilization, and especially for true religion, in the cruelties he was obliged, as he thought, to commit. This would not apply to animals. I have no doubt that

Cortes was **very kind to** them. We have not those after-thoughts about them **that we have** about our fellow men—those after-thoughts which have made **men** severe to their fellows, **sincerely** believing, in **many** instances, that they were ensuring some great **and final good to those whom they persecuted** remorselessly.

I maintain that my dictum **is** substantially right, **that** you have only, by the aid of imagination, to enter fully into what we may reasonably conceive to be the feelings of animals, to be most tender and kind towards them. Even such talk as **we have just** had, which might not appear at first to bear upon the subject, is most useful. Only think of the ways, habits, **and** peculiarities of any creature, and you become tolerant towards it. I will exemplify what I mean. **The horse is** a most timid and nervous animal. By **the way, I** observed that not **one of you was** inclined, in **your** imaginary choice of animal life, to become a horse or any animal that has much **to do** with man. Well, the horse, **as I** said, **is a most timid** and nervous animal. The **moment you have recognized this fact,** you are able, by the **aid of imagination, to enter,** as it were, into its terrors, and you do **not beat a creature merely** because it is afraid.

Cranmer. I come **to what I** always believe in as the main specific for all evils, namely, education. Milverton talks of imagination and makes too much of that, I think. Imagination must have a basis of facts to build upon. Now it is a fact, I believe, as Milverton states, that the horse is a peculiarly nervous and timid animal. It sees a large piece of paper, or an empty sack, or an ungainly-looking shadow in the road, and it takes it to be some dangerous living creature. That ought to be told to people from the first, especially to children. I have never written a book, but it really seems to me that I could write a book, if I possessed the requisite knowledge—

Ellesmere. Well, most of us could do that.

Cranmer. —if I possessed the requisite knowledge, I say. It would be most serviceable in schools, and, indeed, should be a class-book in all schools. I have never read a child's book upon animals which has satisfied me as regards the points I should endeavour to inculcate.

Ellesmere. Upon my word, Cranmer is coming out in a new character. I do not like, however, to hear any man indulge in threats of a painful kind, and I look upon this as a threat, on Cranmer's part, that he will write a book.

Sir Arthur. The words of Cranmer are the words of wisdom, of beaver-like sagacity. I almost think that I will offer to join with Cranmer in writing such a book, only the worst of it is, one is so deficient in facts. One would have to sit under some eminent naturalist for two years. But, most seriously speaking, it might be the just pride of one's life to have written such a book. Going back to my early childhood, I distinctly remember how thoroughly ignorant I was of the very fact Milverton first brought before us, and that I thought my first pony was to be cured of shying by adopting the most severe measures, to force it up to the apparently dangerous object. I am now, I hope, a little wiser as regards the management both of men and horses in this respect; but really I do not see why I should not have been taught this wisdom at a much earlier period.

Milverton. Cranmer is as right as possible; and I accept, with due humility, his correction when he said that imagination must have a basis of facts to build upon.

Ellesmere. There was some practical thing, some plan, which Milverton alluded to in a former conversation. It had relation to horses. He was going to tell us about it when visitors swarmed in the other day.

Milverton. You will only laugh at it, I dare say; but I don't mind telling it you.

One of the great evils in the treatment of animals is, that they are necessarily entrusted to hirelings. Now, the owner of an animal might, in nineteen cases out of twenty, be trusted to exercise what I may call superficial kindness towards it. At any rate, he would avoid intentional unkindness. How he generally errs, as I have shown before, is from the want of sufficient knowledge, just as we err in the sanitary management of our houses and our families, simply from ignorance of what it much concerns us to know.

To lessen this superficial ill-treatment, which perhaps had better be called visible cruelty, would be a considerable gain; and it is only to be acquired by the owner having some means of ascertaining how his animal is treated by his hireling. I would have some method by which (in the case, for example, of cab-driving) there should be a way of communication open between the passenger and the owner. I have often longed to tell the owner how unworthily his agent acts for him. In every hired carriage I would have the means of doing that. A small locked box, in which one could deposit one's complaint, and of which the master alone had the key. would enable one to do so.

I see at once the objections which some of you will make, namely, that complaints would be made from ill-nature or from frivolousness. I don't believe it. The people who ride in hired carriages have too much to do, and are too intent upon their work, to be over-busy in the cause of humanity to animals; but, occasionally, this device might be of great service; and it would be a constant check upon the inhumanity of the driver, if he were an inhuman man. Moreover, it would often touch the owner himself, and be an excuse and an aid to the driver, as when, for instance, an owner sends out an animal which is imperfectly recovered from lameness, or has been badly shod. However, to state the question broadly, to assure the good treatment of animals used in hired carriages, it is most desirable to aim at having some mode of communication with the owner. If this idea is once put before the minds of men strongly, they will find other ways of effecting the purpose besides my way of the little locked box.

Cranmer. I, for one, do not think this suggestion at all a small one; and it seems to me very practical and very practicable.

Ellesmere. It is not bad; but Milverton exaggerates when he says that nineteen out of twenty owners are considerate and humane to their animals, as far as intention goes.

Now I will tell you what is a horror to me, and where I strongly suspect the guilt is upon the owner, and not upon any hireling. You often see a wretched little animal driven at a tremendous pace, with a fearful load behind it of men, women, and children, pleasure-taking, as I suppose. Now, here, I have scarcely a doubt, the owner is the driver. I am sure I don't grudge these people their pleasure; but I lament for the poor animal its want of rest, for you often see that the 'light cart' (light no longer) is one that is probably used for business purposes throughout the week. Now, I am only a Sabbatarian as regards animals. I want the Sunday to be not only the most holy, but the happiest day for all of us. But, as regards rest, animals should be considered first.

We have now dealt handsomely with a practical suggestion. Cranmer approves; so do I; and the others show their consent by a judicious silence.

Let us go into theory a little. There are some fellows amongst us who read books—a weak employment of the human mind, but still a not unfrequent one. Sir Arthur and Milverton are gobblers of books. Cranmer reads books of a cerulean colour to any extent. What do the learned say about the intellect of animals? I have always detested Descartes; and my detestation of him has always been more hearty and complete, by reason of my knowing nothing of his works from personal inspection. But I have been told that he maintained that animals had no feeling. I do not like to make any violent assertion anent the sayings of philosophers; but I think that this is about the most absurd one I ever heard of. I know that we can only detect the *noumenon* from the *phenomenon* (you see I know their terms); but having observed that the pinch which gives me pain and makes me cry out, or be disposed to cry out, produces a similar effect upon the lower animal, I, for my part, require no more proof. But is it true, Sir Arthur, that Descartes asserts this proposition? I should be sorry to dismiss any comfortable detestation from my mind; but justice, of which virtue I am an honourable minister, will compel me to do so, if the accusation is false.

Sir Arthur. It is true. One of his main arguments is, that animals always do things in the same manner without having *learnt* how to do them, and that it would be possible to construct a machine which would have the power of moving about, and of uttering sounds similar to that of an animal.

Mauleverer. I am sorry to hear that the good man talked such nonsense. Before this time I only knew one thing about him; but that had given me a very high opinion of him. I knew that he had taken for his motto '*Qui bene latuit bene vixit.*'

Milverton. Your question, Ellesmere, was about the intellect of animals. I do not know any saying about them which has come home to me more than that of Anaxagoras. I have only learnt of Anaxagoras through Goethe. This is what Goethe says: 'Anaxagoras teaches that all animals have active Reason, but not passive Reason, which, as it were, is the interpreter of the understanding.'*

I don't pretend to understand the last few words; but, as I said before, that saying comes home to my mind with much force. And, if true, it affords a beautiful illustration of the prudence and sparingness of what we call Nature, but which I would rather call Providence. These animals have the powers of reasoning necessary for the guidance of themselves, but not those powers of reflection which would probably be a source of suffering to them.

Ellesmere. Then you think that Fairy has sufficient powers of reasoning to understand how bones are most surely to be obtained in sufficient quantity; and, with that view, resolves to abide here and to be true to her master, whom, I observe, she always reluctantly forsakes even when I know it would be a pleasure to her to go out with another active animal named Ellesmere, in preference to remaining passively with her meat-winner, the inactive Milverton. But that when she stays in his study with him, she in nowise partakes of his reflections upon good and evil, the Finite and the Infinite, the Subjective and the Objective.

---

* Anaragoras lehrt daß alle Thiere die thätige Vernunft haben, aber nicht die leidende, die gleichsam der Dolmetscher des Verstandes ist.

Milverton. No; you confuse affection with reason. Fairy stays with me because she feels (doesn't reason about it) that I am the real, constant, abiding friend, and must be considered first. Wherein she shows the active reason that Anaxagoras speaks of—the reason necessary for her preservation, as it seems to her—is in burying the bones for future meals, when she has for the present satisfied her hunger. I assure you, that the more carefully you work out this saying of Anaxagoras, the more truth you will find in it. You must know that Anaxagoras was one of the greatest philosophers that ever lived.

Ellesmere. Well, we have had enough food for reflection to-day. We have exercised both our active and our passive reason sufficiently; and now I vote we exercise our muscles and take a long walk.

Milverton. No; you must not go away; I have something to read to you—a letter from a Scotch gentleman on the subject of the 'bearing-rein.' Here it is. I shall not give you the beginning of it.

Ellesmere. Yes, do: I should like to hear it all.

Milverton. Don't be so curious. Perhaps there is some comment upon you, saying what a troublesome 'chiel you are to me. I shall only give you what I please.

He says :—

The subject is the use of the 'bearing-rein' for horses. This abominable and useless contrivance is *not used in Scotland* (with perhaps the exception of a permanent check-rein on young, strong, high-spirited carriage horses; and it could not prudently be dispensed with as regards them).

I visit some of the large towns in England twice or thrice a year, and this bearing-rein on heavily laden draught horses is, to my eyes, quite an agonising subject of remark.

I remonstrated with a carter in Manchester, last time I was there, and told him to take it off, and let his horses get their heads down, and their shoulders to the burden. 'Ay, they'd come quick enough down on their knees they would,' was his answer, and no argument would avail.

Now, it came under my observation, that a large number of English horses were brought to Glasgow to work for a railway here, and they had all the bearing-rein on their arrival.

This, however, was an absurdity not to be tolerated by the Scotch carters, who saw at once that the animal was tortured, cribbed and confined in its action, and half the power of the shoulder for drawing was lost—a splendid power! 'His strength is in his neck,' as the old Scripture says.

Well! the horses had been so long used to it, that they could not work without it; but their new masters were not to be baffled, and the next time I saw the horses, they were working with a kind of modified bearing-rein, as follows :—

A longer strap or rope was used, and fastened to the trams of the cart on each side, forming a much less acute angle than the real 'bearing-rein;' and, with this contrivance, the horses were working well, and the look of care and misery was gone from their faces.

Of course this was only an intermediate stage, and ended very shortly in the new horses working altogether without it.

All this was the doing of the Scotch carters themselves, no one interfering with them.

Now, only **suppose that** it should be a result **of** our **conversation,** that bearing-reins should gradually be left **off in** England **as well** as **in** Scotland, what an ample success it would be! **It would** be quite enough to have upon our **tombs, 'He** was one of those men who caused **the** bearing-rein **to** be discontinued.' I should **not wish for** anything more. Sir Arthur would put it beautifully into Latin.

Ellesmere. People do say—I don't say it—that Milverton, with his 'paternal governments,' and things **of** that kind, is very fond **of the bearing-rein** as applied to **human beings.**

Milverton. It is an **outrageous** calumny. There is not anybody **in the world who is more** fond of **personal freedom than I am, and** would protect it more than **I would.** But this is **just** the way of the world. You make a proposition, **guarding it in the most** careful manner, **so that it should not be misconstrued or abused, and** then it is quoted **against** you without any of the guarding part being taken notice of. It really makes one afraid to make any statement whatever. The good Lavater sometimes uses a strong expression when he is especially anxious to recommend to the notice of his hearers some aphoristic saying in which he has the greatest faith. Thus he will exclaim:—

Let the four and twenty elders in Heaven rise before him who, from motives of humanity, can totally suppress an arch, full-pointed, but offensive *bon mot*.

I should like to be able to give equal force to what I am going to say. 'Let the four and twenty elders rise up before him who can repeat an adverse argument fairly.' If you observe, hardly anybody ever does so; and the curious thing is, that people will show this unfairness even in the presence of the man whose argument or statement they misrepresent; and even immediately after he has uttered it. I could credit a man with almost every virtue, if I saw that he represented his opponent's argument fairly and exactly. Of course it requires ability as well as fairness to do this.

Sir Arthur. I agree with you, Milverton, and I would carry the observation further. It is not only that people misrepresent one's statement or one's argument by misquotation—by either omitting something or introducing something—but they adopt a more subtle form of wrong-dealing. They will carry what you have said into a region that it was obviously not meant to apply to—to some subject of a different kind.

Ellesmere. As usual, I protest against abstract sayings. Give us an example.

Sir Arthur. Well, you state something which has reference to political action, in which the passions, the interests, the affections of men are involved; and your adversary immediately applies it to some abstract question of political economy or morals; and he thinks he answers you, because he shows that in this foreign region—foreign to your present region of thought—the saying is not applicable. If I may say so without rudeness, you sometimes do this yourself, Ellesmere. You make a saying appear to be absurd, merely because you will not allow yourself to be bounded by the circle of thought in which your adversary is working.

Milverton. It all comes to this: that the love of intellectual victory is so strong in men that they apply the laws of war to questions of pure thought, and take every advantage of an adversary, being quite indifferent to the research after truth.

Ellesmere. Don't get into a rage, my dear fellow! What does Dr. Blair say about anger?

Milverton. I don't care what Dr. Blair says. Here are a set of people—yourself among them—bound down by the narrowest conventionalities of all kinds; and, because I say that it is necessary, when people live in close contact with one another, as in great towns, that each one should be prevented by some high controlling authority from doing something which directly injures the health and well-being of his neighbours, or when I say that the State, or the municipal authority, should do something for the general good, which no individual can manage to get done for himself, then you talk of my being fond of the bearing-rein for human beings.

Ellesmere. I had no idea that there was a thunderstorm so near!

Milverton. And what a coward you are as regards these conventionalities I spoke of, and which are real bearing-reins to you, and very tight ones too!

Pray, may I ask, Master Ellesmere, are you not, as being one of the most impatient men in the world, often wearied to death at a long dinner?

Ellesmere. I can answer that question honestly—Yes.

Milverton. Did you ever wish to get up, and take a turn about the room, and resume your seat?

Ellesmere. Dozens of times.

Milverton. Did you ever venture to do it?

Ellesmere. No.

Milverton. **I thought so.** There has been but one man in my time brave enough **thus** to risk the danger of his being thought a little eccentric—and who did venture to walk about the room **at dinner-time** when he **was** tired of sitting. **I just mentioned** this trifling thing, which came into my mind **at the** moment, **as** a very slight specimen of the innumerable ties **which** really control human beings, and are their bearing-reins. I don't say that these ties are all of them bad; but what I **do say, is, that** they are infinitely vexatious when compared with **the** control which I would have exercised for the public good.

Ellesmere. Now that Milverton has cooled down a little, I don't mind telling you all a great **secret.** [Here he lowered his voice to a whisper.] **I do not know of anybody,** however agreeable, present **company of course excepted,** whom I do not get tired of as a **neighbour** during a dinner that lasts two or three hours. **As** Dr. Johnson would say, one travels **over** their **minds.**

Milverton. No, one doesn't.

Ellesmere. Nothing **can please him now,** after my unfortunate speech **about the bearing-rein.**

Milverton. It is a most unwarrantable notion, whether Dr. Johnson said it **or not, that one can travel over** other men's minds in **any transient and cursory** manner. The most commonplace person has wild regions—wildernesses it may be—of thought and feeling, which even their most intimate friends hardly ever enter. But I admit that of superficial talk, or even of very good dinner talk, two or three hours is a fearful spell to have with any ordinary human being.

Ellesmere. We are coming round at last.

Milverton. I wish to go back to a part of the subject which we were discussing a few minutes ago. I think I understand Ellesmere. He thinks he understands me, which is much more doubtful. I absolutely foresaw that he would be as sure as possible to introduce the question of the intelligence of animals; and that Descartes' theory, or supposed theory, would have to be discussed. Well, I hunted everywhere to find out what Descartes really said; but I have not come upon any passage which determines the point. I must say that I venture, with all humility, to differ from Sir Arthur, and to declare that I have not found anything in Descartes' writings which justifies the assertion that he maintained that animals have no feeling. His 'Discours de la Méthode' seems to me to point the other way. I did not, therefore, find what I sought for. I have, however, found a treasure. Voltaire once wrote an 'Éloge' upon Descartes. While looking for that, I discovered a short essay of Voltaire's upon animals, which is perfectly admirable. By the way, it is very pious, and would much astonish those persons who suppose that Voltaire had no religion at all.

Sir Arthur. Give us some notion of the essay. I see you have the book in your hand.

Milverton. He begins by an exclamatory sentence, 'What a pity it is! What poverty (of thought) to have said that animals are machines, deprived of knowledge and feeling, always carrying on their labours in the same way, and bringing nothing to perfection!'

Then he gives instances, excellently chosen, of the intelligence of birds. He speaks of the canary learning to sing, and asks this pregnant question: ' N'as-tu pas vu qu'il se méprend et qu'il se corrige?'

He then goes on to describe all the movements of inquietude which a friend might observe in him (Voltaire) when he has lost anything, and of joy when he has found it.

I must now give you his own words:—

'Porte donc le même jugement sur ce chien qui a perdu son maître, qui l'a cherché dans tous les chemins avec des cris douloureux, qui entre dans la maison agité, inquiet, qui descend, qui monte, qui va de chambre en chambre, qui trouve enfin dans son cabinet le maître qu'il aime, et qui lui témoigne sa joie par la douceur de ses cris, par ses sauts, par ses caresses.

'Des barbares saisissent ce chien, qui l'emporte si prodigieusement sur l'homme en amitié ; ils le clouent sur une table, et ils le dissèquent vivant pour te montrer les veines mézaraïques. Tu découvres dans lui tous les mêmes organes de sentiment qui sont dans toi. Réponds-moi, machiniste ; la nature a-t-elle arrangé tous les ressorts du sentiment dans cet animal, afin qu'il ne sente pas ? A-t-il des nerfs pour être impassible ? Ne suppose point cette impertinente contradiction dans la nature.'

He then has a passage which is eminently Voltairian. It occurs in a discussion of the various theories of naturalists, metaphysicians, and theologians, as to the nature of the souls of animals.

'Les âmes des bêtes sont des formes substantielles, a dit Aristote ; après Aristote, l'école arabe ; et après l'école arabe, l'école angélique ; et après l'école angélique, la Sorbonne ; et après la Sorbonne personne au monde.'

You must read the essay for yourselves. It is to be found in the 'Dictionnaire philosophique.'

But the last passage I will read to you now.

'Mais qui fait mouvoir le soufflet des animaux ? Je vous l'ai déjà dit, celui qui fait mouvoir les astres. Le philosophe qui a dit, *Deus est anima brutorum*, avait raison ; mais il devait aller plus loin.'

Sir Arthur. What **wonderfully** lucid language is that of **Voltaire**!

Milverton. **Yes**: how we should all endeavour to imitate that lucidity! I should like never to speak a single sentence, nor to write one, whether as regards the physical handwriting or the composition, which was not perfectly intelligible, as far as language is concerned, to the most ordinary reader. It is not much worth one's while to be greatly disconcerted at anything that may happen to one in this world; but I must confess that I do hate to be misunderstood.

'Misunderstood,' by the way, is an excellent title of an excellent book.

It is a **perfect marvel** to me that several men of our time, men even of genius, **take, comparatively speaking, so** little care to make themselves thoroughly understood. It was my vexation at being misunderstood, that made me a little angry just now with Ellesmere.

Ellesmere. A little **angry**! Well, I know I was very glad that the study table was between us.

Milverton. I did not mean what I then said to be an answer to Ellesmere only, but to other people; perhaps to some among yourselves. I know he adapted his sneer about the bearing-rein to the talk of other people.

One must **stand up** for one's **self** sometimes. I do

declare I have only met with two or three people in the world, John Mill amongst them, who have as much love of, and value for, personal freedom as I have. There are several things which States have interfered with, and still interfere with, in which their interference is, to my mind, not defensible.

When I am in favour of State interference, it is simply upon those occasions whereon, and upon those matters wherein, the common consent, nay, I may almost say the universal consent, of mankind admits that the objects aimed at are for the good of each individual; and more than that, that the individual cannot obtain this good for himself without the controlling power of the State.

Ellesmere. That last sentence is a long one; and before I could assent to it I must have it brought home to my dull mind by instances.

Milverton. Is it not a universal desire on the part of those people who drink water, that it should be pure water? If I showed you that it is impossible, or at least enormously difficult, for human beings, when they are packed close together, to obtain such a primary requisite as pure water without the interference of the State, then I may fairly claim the interference of the State. But if I choose to marry my deceased wife's sister; if I choose to subscribe to a lottery; *if I choose, at one o'clock in the morning, to try and get* a *glass* of *beer, and another person* chooses to keep a house open for the chance of supplying me with it; what are any of these things to you?*

You may say, if you like, that I am fanatical and absurd in my devotion to personal freedom; but, at any rate, do not blame me on both sides, and accuse me of a love for a tight bearing-rein, as applied to men, when you find that I maintain it should be so loose a rein as regards interference with the personal freedom of the people.

Now do not again misunderstand me. I am not fonder of drunkenness and gambling than you are; but I am fond, beyond all measure, of men being allowed to guide themselves and to act upon their own notions of right and wrong in all matters in which the peace and welfare of the community are not directly menaced by the action of individuals.

Now let me take another case, in which control cannot be held to be an unjust interference with personal freedom. There are these millions of people in

---

* Mr. Milverton told me afterwards that he was sorry for having said the words which I have italicised: and he alluded to an article in *Macmillan's Magazine* for November, 1872, which had shown him the immense advantages which had resulted from the Licensing Act—advantages so great as to compensate for the loss of individual freedom in this matter.

a large metropolis like ours. In order that they may circulate freely, and with less danger than, unhappily, they do at present, certain regulations must be made, or ought to be made, by some central authority.

A better instance still is that of erecting a series of barriers, to facilitate the ingress or egress of a great crowd. This is not any interference with private liberty; it is, in fact, a means of creating private liberty. It is not a needless use of a bearing-rein.

In short, when a huge number of human beings are congregated together in a small space, comparatively speaking, you must have regulations to enable even the strong, the wise, and the powerful to have the utmost freedom of action and movement; much more to enable the weak, the ignorant, and the powerless to have the same facilities.

Ellesmere. Well, now do you agree that we have talked enough, and have travelled over a satisfying portion of each other's minds? I know that a little while ago I got into a jungle with some wild beasts in it, which I am very glad to have got out of. And so let us be off for a walk.

He then rose; the party broke up; and there was no more conversation about animals that day.

POSTSCRIPT.—I may mention, that in the course of this conversation Mr. Milverton read the following letter, which he had found in an *Echo* newspaper of last July; the suggestion therein made was highly approved of by himself and the other 'Friends in Council.' I did not introduce it into the conversation on account of its length; but it appears to be well worth considering.

Breaks for Omnibuses.

To the Editor of the 'Echo.'

Sir,—Every afternoon, when I take a peep behind my winkers, I see you, and none but you, fluttering on the top of the lumbering caravan which Destiny has thought good to oblige me to lug along the slippery streets of London, and I often say to myself and to my partner in misery, 'How is it that that 'ere light-weight, as is up to everything, don't pitch into them as is responsible for the needless labour inflicted on us poor horses through the want of drags to stop the 'bus?' Why, you must have seen, times and times, how the collar is pulled nearly over our heads every time we stop, to say nothing of our teeth being crushed, and our necks nearly wrung off (almost all my pals have toothache, but we don't let on that we have it, for it ain't pleasant to have melted lead poured into a tooth, as is how they stop our poor old grinders); and this pretty operation of stopping (I mean the 'bus, not the teeth) is performed a couple of hundred times every day of our lives.

Now, they tell me that there is a gentleman who has power to cause these drags to be put on omnibuses, under the driver's feet, as in Manchester—him, I 'mean, as had little flags stuck on the cabs a short time ago. Suppose he knew that he would be sticking a year or two on our lives by making a stopping drag a condition of a 'bus license, don't you think he would do it? Yes, bless him; but though he is a kind-hearted gentleman, he don't ride on 'buses, and don't know what we have to suffer now. All the drivers know and would bless him if he did so, for I hear them say their arms is pulled off with the stopping (what must our poor mouths be?).

And another thing is, that, if we had drags, less of those stupid bipeds that are continually running before the 'bus, would be injured. I should be rather pleased than otherwise if we now and then ran over a 'bus proprietor; but it goes to my heart when an old person or a little child toddles out into the roadway, and all the people cry out, 'Hoi!' and the driver nearly tugs our heads off, and, do all we can, we can't stop the 'bus in time: it pushes us on before it, and down goes the unfortunate human with a leg broken, or worse! Some of us horses have heard say that there is a society for preventing cruelty to animals (and our treatment is cruelty), but I don't believe it, leastways the society does not, I suppose, ride on 'buses. Now, dear 'Echo,' couldn't you row them up all round, and get us drags, and you shall have our blessing; but here comes Jem to harness me for my day's work, so no more at present from your humble and obedient old

'Bus Horse.

CHAPTER IV.

PREVIOUSLY to recounting the next conversation, I must give some explanations, without which it might not be understood.

We had a short conversation before the one which I am about to relate. In the course of it the general subject was renewed respecting the intelligence and worth of animals. Mr. Milverton and Sir Arthur were required to produce all that they could find upon this subject in the works of authors of renown, and this was to be done in the ensuing ten days. I may mention that Sir John Ellesmere was staying in our house. Mr. Milverton has a way of making everybody work who comes near him; and he employed Sir John, myself, and several other persons, to assist him in his researches, directing them as to what books they should read. Devoting ourselves entirely to this subject (and even the ladies of the house were not allowed any books but those which Mr. Milverton placed in their hands), it is astonishing what a mass of materials we collected.

At length the ten days had elapsed, our friends had re-assembled, and the conversation thus began :—

Ellesmere. Before anything is said on the general subject, I have a quarrel to pick with Milverton. I have now known that man for more years than it is pleasant to enumerate. All this time he has had a book in his library which he must have known it would have been a constant delight to me to read, and which, indeed, using Cranmer's form of words, I should certainly have written if I had possessed the requisite knowledge.

First, though, I must tell you that we have been worked like negro slaves. I have had to read through some of the most detestable books I have ever read in my life—one or two of them in monkish Latin; but when I came to this treasure of a book, I struck work, for it completely fascinated me.

Cranmer. I wonder what it can be. I will make a guess— Search's 'Light of Nature'?

Ellesmere. No. Before, however, I tell you what it is, I would not have you imagine that I have not been of some service in this great enquiry. I have discovered that the saying of Anaxagoras, which Milverton couldn't find, is in Plutarch. Having found that, I think I have done enough.

But now to my own dear book. It is Bishop Berkeley's 'Querist.' You have no idea of the sagacity of that man. Many a conclusion that we take to be the result of modern discovery, is anticipated by him; and then his quaint way of putting everything is so delightful. I must give you some extracts at once, before we get immersed in our subject. He asks:—

Whether reflexion in the better sort might not soon remedy our evils? and whether our real defect be not in a wrong way of thinking?

Whether France and Flanders could have drawn so much money from England for figured silks, lace, and tapestry, if they had not had academies for designing?

Whether our linen manufacture would not find the benefit of this institution; and whether there be anything that makes us fall short of the Dutch in damasks, diapers, and printed linen, but our ignorance in design?

Whether those, who may slight this affair as notional, have sufficiently considered the extensive use of the Art of Design, and its influence in most trades and manufactures, wherein the forms of things are often more regarded than the materials.

Here you see that South Kensington is anticipated.

Again:—

Whether comfortable living doth not produce wants, and wants industry, and industry wealth?

Whether any art or manufacture be so difficult as the making of good laws?

Whether our peers and gentlemen are born legislators? or, whether that faculty be acquired by study and reflexion?

Whether to comprehend the real interest of a people and the means to procure it doth not imply some fund of knowledge, historical, moral, and political, with a faculty of reason improved by learning?

Whether every enemy to learning be not a *Goth*? and whether every such *Goth* among us be not an enemy to the country?

Whether, therefore, it would not be an omen of ill presage, a dreadful phenomenon in the land, if our great men should take it into their heads to deride learning and education?

Whether half the learning and study of these kingdoms is not useless for want of a proper delivery and pronunciation being taught in our schools and colleges?

Whether, in imitation of the Jesuits at Paris, who admit Protestants to study in their College, it may not be right for us also to admit Roman Catholics into our colleges, without obliging them to attend chapel-duties or Catechisms or Divinity lectures?

Whether, as others have supposed an Atlantis or Eutopia, we also may not suppose an Hyperborean Island inhabited by reasonable creatures?

Mauleverer. The extracts are admirable. I decline, however, to say 'Yes' to the last query. Neither in Hyperborean Islands nor anywhere else will you find a country 'inhabited by reasonable creatures.'

I am going to imitate Ellesmere, a practice, however desirable, that I rarely indulge in. Before we separated, the other day, Milverton gave me a book to read, written by a certain Jerome Cardan, whom I had never heard of before, and in this work Milverton was hopeful that I should find something very deep and very significant respecting the nature and character of animals. I did not, however, find anything of the kind; but I found a delicious simile, which will be of use to me for the remainder of my life. There appears to have been a certain Greek, named Theonosto. I wish he had been one of the 'Friends in Council.' I think he would have kept down a good deal of our host's optimist talk, and would invariably have been on my side. He compares the course of our life to water boiling in a cauldron, which, *the more it bubbles up, the less it becomes*, and finally dries up altogether.* It is a lovely simile, and would, if well

---

* Alterum (de quo in Theonosto), qui est secundum virtutem; quique vere optimus est, et totus in potestate nostra: eo magis recordantibus nobis quod (et in Paralipomenis, lib. vi. sect. 2, dictum) vitæ nostræ cursum ferventi in lebete aquæ assimilari; quæ quantumvis intumescat, ubi magis processerit, semper minor sit, et siccatur. Quamobrem optima vita esset, sibi ipsi vivere, conscientiæ, virtuti ac sapientiæ: beate ac feliciter.—Hieronymi Cardani *Arcana Politica*, anno 1635.

attended to, keep people more quiet than they are at present.

Milverton. I really must not allow you to say anything more about Bishop Berkeley, or Jerome Cardan. We have a great deal of hard work before us to-day.

First, I must tell you that there is a most elaborate discussion, or rather two discussions, in Bayle's Dictionary about the reasoning powers of animals. I am very glad that I did not know this until we had made investigation for ourselves, for I might have been tempted to have been satisfied with Bayle, whereas we have found out much that was not known to him. What have you found, Sir Arthur?

Sir Arthur. The most noteworthy passages that I have found are in Seneca. In the first place, he says that animals cannot confer benefits: '*Nec tamen beneficium dant, quod nunquam datur, nisi a volente.*'

In the next place he says that they have no such passion as anger—'*feras irâ carere.*' He admits that they have impulses which create '*rabiem, feritatem, incursum;*' but no such thing as anger.

Ellesmere. I do not wish to say anything rude of that rich man, Nero's tutor; but he is an ass.

Sir Arthur. His doctrine may be very foolish; but it is very interesting to see how early were the seeds of that doctrine implanted which led to

Descartes' theory. **By the way, Milverton,** was I right about **Descartes?**

Milverton. **Yes :. I give up; you were.** Well, now I want to show you something which also anticipates Descartes. There is hardly any writer in the world who had such influence in his own time, and long afterwards, as Thomas Aquinas, the **'Angelic Doctor,'** or the **'Angel of the** Schools,' as he is frequently called.

In his *Summa Totius Theologiæ,* the most tremendous book I believe that ever was written, he takes up the cause against animals. **Here** is one of his propositions: *Animalia bruta non delectantur visibilibus, odoribus, et sonis, nisi in ordine ad sustentationem naturæ.*

Ellesmere. What an absurdity! **Why, with only** my little knowledge of animals, I could mention **to you a dozen** creatures of my acquaintance who have had the **keenest** delight in music for music's sake. I knew a cat who not only delighted in music, but had the nicest perception **as to** who was the best musician in the room (at least pussy's opinion and mine always agreed), **and** she would distinguish her musical favourite by **extreme marks of** approbation and applause.

What a deadly thing it is when once a man has got **a theory** or a doctrine into his head to which he must make everything bend! Or, to put it after the fashion of Berkeley, Whether a man ever makes a greater fool of his understanding than when he adopts some theory, moral, metaphysical, or theological, which he thinks will explain all morality, theology, or metaphysics, and which admits of no exceptions?

Cranmer. I foresee that we shall be inundated with queries from Ellesmere, put in the Berkeleyan form; and that we shall come to dread the word 'Whether' as much as Ellesmere dreads the words 'That reminds me.'

Ellesmere. I flatter myself that I know exactly when a man becomes a bore. I wish that other people were as sensitive upon that point as I am.

But I must tell you something which Milverton said, in the course of these ten days that we have had together, touching this subject, and which really was not a foolish remark. He was telling me how these theories arose which were so inimical to animals: as, for instance, that certain philosophers had come to the conclusion that there were only two things in the world, namely matter and soul. Now they did not like to say that the brutes had soul, for there was but one kind of soul, and that was possessed by man. They did not even like to admit of any admixture of soul and matter, which would create an inferior kind of soul. Consequently, having the full courage o their opinions, they resolved to say that brutes were only compounded of matter, and did not indulge in feelings.

All this, however, was merely information that he gave me. Now for his general remark. He said, 'You may almost always detect the severe, hard, cruel man, by his dislike, even in matters of the intellect, to admit of exceptions. Never, he added, come under the power of a man, if you can help it, whom you perceive to have an especial aversion to admit exceptions to any theory or any rule he has once laid down. That man will be a very hard man to deal with.'

This discourse pleased me, for I am a lover of exceptions, and am consequently rather a soft and gentle person to deal with.

Cranmer. Objections are not exceptions; or, to put the matter more plainly, captious objections are not necessarily judicious exceptions.

Ellesmere. This man has been very angry with me ever since I said that he would make a good tortoise in a future state of existence; and he has been anxious to show that there are creatures of the tortoise species endowed with much force and vivacity—snapping-turtles, for example.

Milverton. This digression is very gratifying to me, as showing that I once made a sensible remark; but may I entreat you, Ellesmere, as a reward for this exertion on my part, to allow me to go on with my subject? I will at once furnish you with a singular contrast to the words of the severe schoolman, Thomas Aquinas, in the pretty sayings of dear old Montaigne. He says:—

If it be justice to render to everyone their due, the beasts that serve, love, and defend their benefactors, and that pursue and fall upon strangers and those who offend them, do in this represent a certain air of our justice: as *also in observing a very equitable equality in the distribution of what they have to their young*; and as to friendship, they have it without comparison more lively and constant than men have. There are inclinations of affection, which sometimes spring up in us without the consultation of reason, and by a fortuitous temerity, which others call sympathy: of which beasts are capable as we: we see horses take such an acquaintance with one another, that we have much ado to make them eat or travel when separated; we observe them to fancy a particular colour in those of their own kind, and, where they meet it, run to it with great joy and demonstrations of good-will, and have a dislike and hatred for some other colour. Animals have choice, as well as we, in their amours; neither are they exempt from our jealousies and implacable malice.

This occurs in his 'Apology for Raimond de Sebonde.' In another essay, **that on cruelty**, he says:—

I hardly ever take any beast or bird alive that I do not presently turn loose. Pythagoras bought them, and fishes, of huntsmen, fowlers, and fishermen, to do the same. Those natures that are sanguinary towards beasts, discover a natural propensity to cruelty. After they had accustomed themselves at Rome to spectacles of the slaughter of animals, they proceeded to those of the slaughter of men, the gladiators. Nature has herself, I doubt, imprinted in man a kind of instinct to inhumanity; nobody takes pleasure in seeing beasts play and caress one another, but everyone is delighted with seeing them dismember and tear one another to pieces. And that I may not be laughed at for the sympathy I have with them, Theology itself enjoins us some favour in their behalf; and considering that one and the same master has lodged us together in this palace for his service, and that they as well as we are of his family, it has reason to enjoin us some affection and regard to them.

Ellesmere. How pleased Montaigne would have been with Horace Walpole (you needn't remark, Cranmer, that this is 'impossible,' because Horace Walpole lived some years after Montaigne); how pleased he would have been, I say, if he could have read that letter of Walpole's to his friend, Lord Strafford, in which, if I recollect rightly, he commemorates the loss of a dog, and says something of this kind, 'If I could have a friend possessing such fidelity, I should not at all mind his having two additional legs.'

Milverton. From Montaigne I pass to Petrarch, who, in his 'View of Human Life,' has the following charming passage :—

Leave all animals to their proper places and their proper uses; those that are wild to the woods, and the direction of Providence for their haunts and their destination; and domestic animals to those whose wide grounds and fields can with wholesome and true care nourish them for thy table, and coop them not up to fret, and waste, and scrape, and litter in thy small inclosures or narrow courts. Suffer also the little birds to live in the open air; there to feed, to multiply, to sing, to stretch out their wings, and smooth their little breasts in joy; and ye little babes, as saith Solomon, turn ye at my rebuke, bring them not to you to pine and die in your domestic prisons; but rather go to them, stretch forth your slothful minds unto Heaven, and join in the full choir of praise to that Power who created the birds of the air, and the fishes of the sea, and man to govern them all, wisely and kindly, for his good.*

From Petrarch I pass to Fuller, who, in describing the good Master, says:—

He is tender of his servant in his sicknesse and age. If crippled in his service, his house is his hospital; yet how many throw away those dry bones, out of the which themselves have suck'd the marrow? It is as usuall to see a young serving-man an old beggar, as to see a light horse first from the great saddle of a nobleman to come to the hackney-coach, and at last die drawing a carre. But the good master is not like the cruell hunter in the fable, who beats his old dogge, because his toothlesse mouth let go

---

* Mrs. Dobson's Translation, 1797.

the game : he rather imitates the noble nature of our Prince Henry, who took order for the keeping of an old English mastiffe, which had made a lion runne away. Good reason, good service in age should be rewarded. And well may masters consider how easie a transposition it had been for God, to have made him to mount into the saddle that holds the stirrup, and him to sit down at the table, who stands by with a trencher.

The above is not a passage directly bearing upon the subject of animals, but it is all the more striking on that ground, as it shows how completely Fuller identified the good service of an animal with the good service of a serving-man.

Everyone knows the sayings of St. Francis of Assisi respecting animals, and how he called the birds his 'dear brothers and sisters.' A modern writer of the saint's life thus describes the love of the saint for all created beings :—

L'amour de saint François pour la nature n'est pas moins célèbre dans les légendes que son inépuisable mansuétude. Et ce qu'il a de particulier, c'est qu'il ne se restreint pas dans son cœur à un être particulier et à quelques moments d'effusion, il s'étend à tout ce qui existe et anime pour ainsi dire chaque instant de sa vie. Ici, nous le voyons se détourner pour ne pas écraser le ver du chemin. Là, assis près d'un figuier, il appelle une cigale et lui commande de louer Dieu ; la cigale obéit, vole sur sa main, et tous les jours elle venait visiter le patriarche des pauvres et lui élever le cœur par ses chants. L'hiver venu, il avait une grande crainte que les abeilles ne mourussent de froid, et il leur faisait apporter du miel et du vin.*

I think we may assume that the tenets of their great founder respecting animals, were held in reverence by the Franciscan monks. One instance of the kind I remember to have seen noticed by Humboldt. He mentions that a Franciscan, who had accompanied him through some of the most difficult country in South America, used to say, when apprehensive of a storm at night, 'May Heaven grant a quiet night both to us and to the wild beasts of the forest!' Humboldt had mentioned that a storm at night creates great terror and confusion among these wild beasts.

Again, the greatest poets in all ages have been great admirers of animals, and their sayings would form a code of tenderness for these our fellow-creatures. Throughout Shakespeare's writings, for instance, you can detect the love that he had for animals.

Ellesmere. I don't think much of that. Shakespeare understood everybody and everything, and accordingly liked everybody and everything.

---

* Saint François d'Assise par Frédéric Morin. Paris, 1853.

Cranmer. Certainly. It is clear he understood all about lawyers, and yet he never speaks unkindly even of them.

Ellesmere. How sharp Cranmer is becoming; but I wish he would not always whet his wit on me.

Milverton. Now I am going to quote a very remarkable passage or series of passages; and what is remarkable in them is this—that somewhat of the same idea as that put forth by Seneca, by Thomas Aquinas, by Descartes (by Leibnitz also, if I recollect rightly), has evidently some hold upon a comparatively modern poet—a very religious man. You see the same course of thought; and you feel that if this man had lived in the times of Thomas Aquinas, he would have agreed with the 'Angelic Doctor.' It is from Young's 'Night Thoughts.' He is proving that the soul of man is immortal. Young, I am afraid, is very little read now, but it seems to me that he is a great poet.

> If such is man's allotment, what is heav'n?
> Or own the soul immortal, or blaspheme.
> Or own the soul immortal, or invert
> All order. Go, mock-majesty! go, man!
> And bow to thy superiors of the stall;
> Thro' every scene of sense superior far:
> They graze the turf untill'd, they drink the stream
> Unbrew'd, and ever full, and unembitter'd
> With doubts, fears, fruitless hopes, regrets, despairs;

Mankind's peculiar, Reason's precious dow'r !
No foreign clime they ransack for their **robes**;
Nor brothers cite to the litigious **bar**;
Their good is good entire, unmix'd, unmarr'd;
They find a paradise in every field,
On boughs forbidden where no curses hang:
Their ill, no more than strikes the sense, unstretch'd
By previous dread, or murmur in the rear;
When the worst **comes,** it comes unfear'd; one stroke
Begins and ends their woe: they die but once;
Blest, incommunicable privilege! for which
Proud man, who rules the globe, and reads the stars,
Philosopher, or hero, sighs in vain.

Then, having said so much for brute life, he restores man to his supremacy, of course **inferring that there is** no future existence for brutes :—

Account for this prerogative in brutes.
No day, no glimpse of day, to solve the knot,
But what beams on it from eternity.
O sole and sweet solution! That unties
The difficult and softens the severe,
The cloud on Nature's beauteous face dispels,
Restores **bright order, casts** he brute beneath,
And re-enthrones us in supremacy
Of joy, ev'n here; admit immortal life,
And virtue is knight-errantry no more;
Each virtue brings in hand a golden dow'r.
Far richer in reversion: hope exults;
And, tho' much bitter in our cup **is thrown,**
Predominates, and gives the taste of heav'n.

The philosophers whom we **have** quoted do not trouble themselves about the lot in this life assigned to the animal creation. Several of them, indeed, conclude that the animals have no such lot, inasmuch as they are mere *phenomena* and *simulacra*; but Young could not do that, and would not wish to do that. He gives them a certain glory and a certain fulness of happiness in this life. He is not perplexed by the thought which presses upon the severe thinker, Bayle, and which, I doubt not, has been shared by many a thinker, especially in modern times.

'Les actions des bêtes,' says Bayle, 'sont peut-être un des plus profonds abîmes sur quoi notre raison se puisse exercer ; et je suis surpris que si peu de gens s'en aperçoivent.'

Sir Arthur. Without pretending to be a severe thinker, I entirely accord with what Bayle says upon the subject.

Milverton. Most appropriately does our friend Johnson now hand me a book in which there is a quotation from Dr. Arnold. He says :—

It should seem as if the primitive Christians, by laying so much stress upon a future life in contradistinction to this life, and placing the lower creatures out of the pale of hope, placed them at the same time out of the pale of sympathy, and thus laid the foundation for this utter disregard of animals in the light of our fellow creatures. Their definition of virtue was the same as Paley's—that it was good performed for the sake of ensuring everlasting happiness—which of course excluded all the so-called brute creatures. Kind, loving, submissive, conscientious, much-enduring, we know them to be; but because we deprive them of all stake in the future, because they have no selfish, calculated aim, these are not virtues; yet if we say 'a *vicious* horse,' why not say 'a *virtuous* horse'?

I think this comes in well after what we have heard from Young's 'Night Thoughts.' Johnson found it in a work of Mrs. Jameson's,* who has written very well upon the subject. She also gives the following passage from Jeremy Bentham :—

The day may come when the rest of the animal creation may acquire those rights which never could have been withheld from them but by the hand of tyranny. It may come one day to be recognized that the number of legs, the villosity of the skin, or the termination of the *os sacrum*, are reasons insufficient for abandoning a sensitive being to the caprice of a tormentor. What else is it that should trace the insuperable line? Is it the faculty of reason, or perhaps the faculty of discourse? But a full-grown horse or dog is beyond comparison a more rational as well as a more conversable animal than an infant of a day, a week, or even a month old. But suppose the case were otherwise, what would it avail? The question is not 'can they reason? nor 'can they speak?' but 'can they suffer?'

I quote that passage because the final sentence is exactly like one that I gave you the other day from Voltaire's essay, and it really contains the gist of the

---

* Mrs. Jameson's 'Commonplace Book of Thoughts, Memories, and Fancies.' London, 1854.

subject. Johnson, I see, wishes to read us something from an author who has written very well upon this subject.

Johnson. It is from Grindon's 'Life: its Nature, Varieties, and Phenomena.' He says:—

The doctrine of the immortality of brutes is an exceedingly ancient one. The Indian, whose blissful heaven consists of exhaustless hunting-grounds, does but reflect from the forests of the West what is thousands of years old in the Odyssey:—'After him I beheld vast Orion, hunting, in the meadows of asphodel, beasts which he had killed in the desert mountains, having a brazen club in his hands, for ever unbroken.' Virgil, in his sixth book, enumerates animals seen by Æneas in the kingdom of Pluto; Hercules, in Theocritus, finishes the narration of his great exploit of slaying the Nemæan lion by saying that 'Hades received a monster soul.' 'As brutes,' says Richard Dean, Curate of Middleton in 1768, 'have accompanied man in all his capital calamities (as deluges, famines, and pestilences), so will they attend him in his final deliverance.' Dr. Barclay ('Inquiry,' &c., p. 339) pleads that, for aught we know, brutes *may* be immortal, 'reserved, as forming many of the accustomed links in the chain of being, and by preserving the chain entire, contribute, in the future state, as they do here, to the general beauty and variety of the universe, a source, not only of sublime, but of perpetual delight.'

Milverton. I must now give you a bit from an old translation of 'Plutarch's Lives.'* It occurs in his life

---

* 'Plutarch's Lives,' translated by Sir Thomas North, Knight. London, 1612.

of that hard man, Marcus Cato, whom the biographer is blaming :—

For, we see, gentlenesse goeth further than justice. For nature teacheth us to use justice only to slaves, but gentlenesse sometimes is shewed unto bruite beasts : and that cometh from the very fountaine and spring of all courtesie and humanity, which should never dry up in any man living. For to say truly, to keep cast horses spoiled in our service, and dogs also, not onely when they are whelpes, but when they be old, be even tokens of love and kindness. As the Athenians made a law, when they builded their temple called *Hecatompedon*; that they should suffer the moyles and mulets that did service in their carriages about the building of the same, to graze everywhere, without let or trouble of any man.

Ellesmere. I like the idea of using courtesy to animals. They are very appreciative of politeness, and observant of the reverse. They like to be laughed with, but have a great objection to be laughed at.

I think I mentioned to you at the beginning of our conversation that I had not been a useless searcher. It was only when I came upon 'Berkeley's Querist' that I ceased to act as a pointer dog for Milverton. Now in Hudibras——

Cranmer. I believe no other human being but Sir John would ever have thought of looking in Hudibras.

Ellesmere. In Hudibras there is a sly hit at the

sayings of the philosophers, which is exceedingly well put :—

> They rode, but authors having not
> Determin'd whether pace or trot,
> That is to say, whether tollutation,
> As they do term't, or succussation,
> We leave it, and go on, as now
> Suppose they did, no matter how ;
> Yet some, from subtle hints have got
> Mysterious lights it was a trot :
> But let that pass ; they now begun
> To spur their living engines on :
> For as whipp'd tops and bandy'd balls,
> The learned hold, are animals ; *
> So horses they affirm to be
> Mere engines made by geometry,
> And were invented first from engines,
> As Indian Britons were from penguins.

Observe the sly way in which he insinuates that the

[* The atomic philosophers Democritus, Epicurus, &c., and some of the moderns likewise, as Descartes, Hobbes, and others, will not allow animals to have a spontaneous and living principle in them, but maintain that life and sensation are generated out of matter, from the contexture of atoms, or some peculiar composition of magnitudes, figures, sites and motions, and consequently that they are nothing but local motion and mechanism. By such argument tops and balls seem as much animated as dogs and horses. Mr. Boyle, in his 'Experiments,' printed in 1659, observes how like animals (men excepted) are to mechanical instruments.—*Note in the Rev. Dr. Nash's edition of Hudibras.*]

same arguments which made animals to be machines, would make whipped tops into animals.

Cranmer. Did you suggest Hudibras to Ellesmere, Milverton?

Ellesmere. Not he: nothing so sensible. He always gave me books that required a constant and humiliating reference to dictionaries. After a certain age, one becomes a little tired of looking into dictionaries.

I wish you could have seen and heard the insinuating way in which he would endeavour to inflict the most dreadful books upon me. On a fine morning, after breakfast, when I am most radiant, and anybody can get me to do almost anything, he would come up to me and say, 'My dear Ellesmere, here is the *Tractatus Theologico-Politicus,* a work of Spinoza's, one of his best, as the critics say : I dare say you have not read much of Spinoza' (just as if I had ever read any of him !) 'You will find it very interesting ; and there is sure to be something in it relating to our subject.'

Well ! what can one say to a man in his own house when one has just been eating his bread and butter? I felt that I was looking very blank and disheartened. I looked out of the window, too, but the hint was not taken; and so I went away sorrowfully with my *Tractatus* to my own room. I read twenty pages of it, but can assure you that there were no three consecutive sentences of which I could thoroughly make out the meaning. So I cast the book down; and, from the love of contrast, took up Hudibras; and there, as you see, I had what boys call 'a good find.'

Milverton. Ellesmere showed me that passage from Hudibras, and I was at first much puzzled to understand what Hudibras meant by using the word 'Penguins.' A passage, however, in a work of George Lewes's, a writer on biology of wonderful clearness and truthfulness, explains the whole thing. And it is well worth quoting for its own sake.

It is curious to find men in all ages laying so much stress on a very unimportant peculiarity, and making man's supremacy to consist in a power of gazing upwards, which is shared by every goose that waddles across his path.

'L'homme élève un front noble et regarde les cieux,'

says Louis Racine, in imitation of Ovid's well-known lines:

'Pronaque cum spectent animalia cætera terram,
 Os homini sublime dedit, cœlumque tueri
 Jussit.'

Galen justly ridicules this notion; it is, he says, refuted by the fact that there are fish which always have their eyes directed towards the heavens, and that man can only direct his eyes upwards by bending back his head. As to the erect position, no one till Isidore Geoffroy St. Hilaire thought of the familiar fact that many birds, such as penguins, have the vertical attitude, and some mammals—such as the gerboa and kangaroo—approach it very closely. If the attitude of man is more perfectly erect, this is but a question of degree, not worth making a cardinal distinction.

Sir Arthur. I am sorry to say, Milverton, that you have still failed in explaining the passage in Hudibras. Dr. Nash justly says, that when Hudibras mentioned penguins, he meant to ridicule his friend, Selden, who had a notion that the Welsh had discovered America, on account of the similarity of some words in the two languages. 'Penguin, the name of a bird with a white head in America, in British signifies a white rock;' but I am very glad that you quoted the passage from Lewes, because it gives me the opportunity of mentioning other foolish notions which men have taken up in order to show their infinite superiority to animals. The great Aristotle claims superiority for man because he is ticklish, and the other animals are not, and he puts forth this amazing statement—'Man, alone, presents the phenomenon of heart-beating, because he, alone, is moved by hope and expectation of what is coming.'

Milverton. Yes; and Lewes makes this good remark upon the passage:—'One would fancy that Aristotle had never held a bird in his hand.'

Ellesmere. Really, I am thankful that I am not a philosopher. If I were, I should be so ashamed of the sayings of a great many of my brother philosophers.

Milverton. Now to return to the *Tractatus Theologico-Politicus.* Owing to Ellesmere's idleness, I had to look through this book, of the difficulties in which he speaks in **such** exaggerated terms, and I found one passage which clearly indicates that Spinoza did not adopt the views of Descartes about animals, for he contrasts animals with automata:—*Non, inquam, finis reipublicæ* **est** *homines* **ex rationalibus** *bestias* **vel** *automata facere, sed contra ut* **corum mens** *et corpus tuto suis functionibus fungantur, et* **ipsi** *liberâ ratione utantur, et ne odio, irâ, vel dolo certent,* **nec** *animo iniquo invicem ferantur.**

Ellesmere. I trust we shall now have a little respite from questionable **Latin.** But we have not yet heard much from Sir Arthur. I expected him to have brought a **fearful** amount of learned **grist to** the mill!

Sir Arthur. I **had** prepared **a** great many passages from the Latin poets, to show their feelings of kindliness towards animals.

Ellesmere. I do think we may **consider** those as read.

Sir Arthur. I **told** you, Ellesmere, that **I** *had* prepared; meaning **to convey to** your mind that **I did** not intend **to inflict these** passages upon you. After all, they **prove but little.** One **knows** beforehand that **sensitive men, such as poets are, would be** sure to take a kindly view of **the** animal **creation;** and, moreover, **that** their habits **of observation** would make them **notice** the ways of animals, **and feel a** regard for them.

---

* Cap. xx. p. 227.

Milverton is quite right when he says, 'You have only to observe their habits to become fond of them;' and that, by the way, is the reason why I venture to differ a little from Milverton and Petrarch in any protest they make against the keeping of pets.

I must give you one quotation from Lucretius. It is not that it bears closely upon the subject, but because it sounds to me so grand. I particularly admire the word *bacchatur*, which, as you will hear, occurs in the passage:—

> At novitas mundi nec frigora dura ciebat,
> Nec nimios æstus, nec magnis viribus auras :
> Omnia enim pariter crescunt, et robora sumunt.
> Quare etiam atque etiam maternum nomen adepta
> Terra tenet meritò, quoniam genus ipsa creavit
> Humanum, atque animal propè certo tempore fudit
> Omne, quod in magnis bacchatur montibu' passim,
> Aëriasque simul volucreis variantibu' formis.

This is Munro's translation:—

But the early age of the world gave forth neither severe cold, nor extraordinary heat, nor winds of impetuous violence. For all these alike increase and acquire strength *by time.*

For which cause, *I say* again and again, the earth has justly acquired, *and justly* retains, the name of MOTHER, since she herself brought forth the race of men, and produced, at *this* certain time, almost every kind of animal which exults over the vast mountains, and the birds of the air, at the same period, with *all their* varied forms.

I said just now that the poets were sure to speak kindly of animals; but when a man was a poet and a Pythagorean too, it was doubly incumbent upon him to promote a merciful feeling towards animals. I find that Xenophanes, as translated by Mr. Edwin Arnold in his 'Poets of Greece,' says:—

> Going abroad, he saw one day a hound was beaten sore;
> Whereat his heart grew pitiful: 'Now beat the hound no more!
> Give o'er thy cruel blows,' he cried; 'a man's soul verily
> Is lodged in that same crouching beast—I know him by the cry.'*

Ellesmere. I think we have got off very easily from the weight of Latin poetry which Sir Arthur would have poured out upon us with that rotund utterance which scholars delight in, dwelling upon each word as if it were a pleasure and an honour to utter it.

Milverton. I am now going to take you far away from the Romans to David Hume. In his essays, there is a section entitled 'Of the Reason of Animals.' He gives instances of the sagacity of animals, and says:—

A horse that has been accustomed to the field, becomes acquainted with the proper height which he can leap, and

---

* Καὶ ποτέ μιν στυφελιζομένου σκύλακος παριόντα
φασὶν ἐποικτεῖραι καὶ τόδε φάσθαι ἔπος·
Παῦσαι μηδὲ ῥάπιζ', ἐπεὶ ἦ φίλου ἀνέρος ἐστὶ
ψυχὴ τὴν ἔγνων φθεγξαμένης ἀΐων.

will never attempt what exceeds his force and ability. An old greyhound will trust the more fatiguing part of the chase to the younger, and will place himself so as to meet the hare in her doubles; nor are the conjectures which he forms on this occasion founded in anything but his observation and experience.

But then Hume will not allow you to say that this sagacity results from a conclusion that 'like events must follow like objects, and that the course of nature must always be regular in its operations.'

On the contrary, he says :—

Animals, therefore, are not guided in these inferences by reasoning: neither are children; neither are the generality of mankind in their ordinary actions and conclusions; neither are philosophers themselves, who, in all the active parts of life, are in the main the same with the vulgar, and are governed by the same maxims. Nature must have provided some other principle, of more ready and more general use and application; nor can an operation of such immense consequence in life as that of inferring effects from causes, be trusted to the uncertain process of reasoning and argumentation.

Now this appears to me to be a most sensible way of looking at the matter. I confess I had never thought of it in this way before. What Hume means by 'some other principle' may be seen from the following sentence which occurs at another part of the essay :—

It seems evident that animals, as well as men, learn many things from experience, and infer that the same events will always follow from the same causes. By this principle they become acquainted with the more obvious properties of external objects, and gradually, from their birth, treasure up a knowledge of the nature of fire, water, earth, stones, heights, depths, &c., and the effects which result from their operation.

Ellesmere. I really think this is excellent. What a relief it is to come from the wiredrawn nonsense of Seneca, Thomas Aquinas, and Descartes, to the broad common sense of this thoughtful Scotchman! Now Milverton never gave me such an author as Hume to study. I should have been sure to have noted these passages.

Milverton. Now shall I leave off? I am afraid of wearying you with these quotations; but I am desirous of bringing before you a great body of evidence, to show what have been the thoughts of remarkable men, for many ages, respecting the nature and treatment of animals. And I have still a good deal more to bring forward on this subject.

Ellesmere. I am for leaving off. My mind is one that soon becomes tired of much contemplation. I decline, for instance, to see more than six pictures at a time in any gallery. I shall give much more attention to the subject if you will humour my proneness to intellectual fatigue.

Sir Arthur. I confess I should wish to go on.

I like to have a great quantity of evidence brought before me at once ; and I have no doubt, Cranmer would too ; but we must humour this spoilt child, Ellesmere, and so you, Milverton, must stop for the present.

CHAPTER V.

THE conversation was thus resumed on the following morning :—

Cranmer. Have you not often remarked that when one has any particular subject in one's thoughts, that subject seems to be brought up on all occasions? It seems almost as if it was in the air that surrounded you. Of course, this is an utter delusion. It is merely that one's attention is aroused to everything that is said by anybody bearing upon the subject.

Ellesmere. Observe, how afraid Cranmer is of admitting, that there is anything magical or mysterious in the world. Now I, being of a romantic and poetical nature, believe that Cranmer is so deeply impressed with the subject we are discussing, that a certain sense of it radiates from him unconsciously, and affects all the people who come within the sphere of influence of that magnetic mind. Now, go on, Cranmer.

Cranmer. Well, when I left you the other day, I was thinking over our subject, when suddenly a gentleman in the train said that he had been witness to a great instance of cruelty to living creatures, which had been brought under his notice at a station on the railway. He said that he was a member of that excellent Society for the Prevention of Cruelty to Animals, and had communicated with them upon the subject. I asked him to give me a written statement of the facts he had observed. Here it is, I will read it to you—

At seven o'clock on the night of the 7th inst. thirty-six boxes of live geese arrived at Waterloo Station from St. Malo, viâ Southampton, consigned to Mr. ——, Leadenhall Market. Each box appeared to be three feet four inches long, two feet wide, and sixteen inches deep; and all were made of rough jagged-edged deal planks, left with openings between each plank at the top and sides. In every box, so far as I could tell, from nine to twelve geese were huddled together so closely that none could move except by trampling one over another; or by getting a neck, head, or wing out of one of the openings. Some of the geese were screaming, many were lying down with heads and necks extended, seemingly quite exhausted; several were dead. I could count three, but believe there must have been more, the boxes being so placed in a mass on the platform that I could only examine closely those that were outermost.

It was painful to see heads, necks, and wings protruding from the boxes, so firmly fixed in openings that moderate force could not remove them. But it was still more painful to see how eagerly those geese which could get their heads out freely drank up some water the porters sprinkled on the boxes. The geese were so crowded together it would have been impossible to give them either food or water in the boxes, and I greatly fear they must have been left in them all night, as there was no preparation for their removal when I left at twenty minutes past eleven. I could not learn how long they had been in the boxes.

Now, isn't this shocking!

Mrs. Milverton. Abominable.

Lady Ellesmere. It is all very well to speak scornfully of us women; but I do seriously think, that if we had more voice in the management of affairs, these things which you call minor matters would be very differently managed. Say what you like, we are more humane. To feel deeply for these creatures is just the sort of thing that many men would laugh at, and call it sentimental, and a spurious kind of humanity, merely because the creatures are not big.

Ellesmere. Please don't include me among the naughty men whom you condemn. I detest cruelty to any creature, whether big or little. But now, Milverton, comes the difficult question of what ought to be done in this case. You can't well attempt to legislate upon the subject of geese.

Sir Arthur. I am not prepared to admit that. We do legislate about other birds.

Milverton. Yes; there would come that foolish ridicule which the Scripture likens to 'the crackling of thorns under a pot;' and I doubt, Sir Arthur, whether even your eloquence, backed by Cranmer's skill and perseverance, could get a Bill on the subject of geese through Parliament.

First, however, I must ask Cranmer whether he will admit that, according to the supreme laws of political economy, we may interfere.

Cranmer. Of course I do. It is really a shame always to fix upon me some extravagant interpretation of these laws.

Milverton. Well, then, I will tell you how I should propose to deal with such a matter. I, myself, think nothing small that is inhuman, but I own to you that I have a great fear of the damage that ridicule might do to any proposed legislation, which should have for its object, *directly,* the improvement of transit for the smaller creatures. I know as well as possible, that even our present discussions would be carped at on account of their being chiefly devoted to humanity to animals. People would say, 'Are there not enormous grievances which affect the higher creatures of creation, and had you not better attack these grievances first?' Well, in former days we used to discuss the subject of slavery, and I am sure we have discussed enough the subject of war. On the present occasion we forsake mankind for the moment, and endeavour to say something for those poor creatures that can say nothing for themselves.

Ellesmere. But what is your plan, Milverton? How would you legislate so as to check the inhumanity in question?

Milverton. I want the whole subject of the transit of living creatures to be reconsidered. Nothing in this world is an unmixed benefit. The increased facility of locomotion by railway has introduced new elements of difficulty into the whole question.

How I should endeavour to meet this particular case, is by the adoption of some general rules, similar to those which have been introduced into the Passengers' Act, 1855, and subsequent Acts, with relation to the transit of human beings. Don't let us talk about ducks, or geese, or any such small fry; but let us contend for a provision of this kind—that in all cases of transit of living creatures a certain space should be allowed, bearing some proportion to the size of the creatures respectively.

Lady *Ellesmere.* Such a provision, although no doubt, a very good thing, will not alone satisfy me, Leonard. The form given to the means of conveyance must also be considered. Now, in this very case, it is evident that these poor birds suffered greatly on account of that form being most inappropriate. What a cargo of animal suffering this was, that your friend in the railway spoke of, Mr. Cranmer! It horrifies one to think sometimes of what other creatures are suffering while we are sleeping. Imagine their nights of suffering! Think only of one thing, the want of water!! Now, Leonard, if you have any regulations for the transit of animals, due supply of water must, indeed, be one of them.

Ellesmere. I wish we could delegate to women some of this work. I should approve of this more than of their contending with me in the Queen's Bench, which at last, I suppose, they will insist upon.

Milverton. I am going to say something rather rough and strong; but I must say it. People talk of our being damned for this and that, using the word damnation very freely. I sometimes think, when I meet with, or hear of, these cases of cruelty, that what we men run a risk of being damned for, is for our barbarity to these creatures who have been given into our complete dominion, and for our conduct to whom we shall be fearfully answerable.

Sir Arthur. Now, Milverton, after this long episode, I think we shall return with greater zest to your quotations from sundry authors, seeing that from this instance, which Cranmer has brought before us, there is pressing need for the inculcation of humanity to animals from whatever source we can derive it.

Milverton. The next author from whom I shall quote is Lord Brougham. His ' Dialogues on Instinct' is a remarkable book if only for its research, considering that it was **written in** 1837, when he was an active politician.

Althorp. I can well suppose a difference merely in degree, sufficient to explain any diversity of condition or result. We have only to compare individual men together to perceive this. It is admitted that reason, nay, that the power of forming abstract ideas, as well as drawing inferences from premises, is possessed by persons whom yet you shall in vain attempt to teach the simplest mathematical demonstration. Then their faculties differ only in degree from those by which Pascal learnt geometry without a master or a book, and Newton discovered Fluxions, and Lagrange and Euler the Calculus of Variations. It may truly be said, that there is no difference in kind which could make a greater diversity in the result.

Brougham. It may indeed be truly so said; but it may also be added, that there is not a greater difference, call it in kind or in degree, between the person whose obtuseness you have supposed and a sagacious retriever, or a clever ape, than between the great mathematicians you have named and that same person. Locke, whose calmness of understanding was equal to his sagacity, and never allowed his judgment to be warped by prejudice, or carried away by fancy and feelings, seems to have held this opinion, and indeed to have allowed some reason to animals. 'There are some brutes,' he observes, 'that seem to have as much knowledge and reason as some that are called men;' and he goes on to say, that there is such a connexion between the animal and vegetable kingdom, as makes the difference scarcely perceptible between the lowest of the one and the highest of the other.*

---

* *Dialogues on Instinct*, p. 141, edit. of 1844.

I must, however, take a strong objection to the course of his argument, or rather, to one of the main facts on which it rests. I do not believe that anyone who has the power of forming abstract ideas, as well as drawing inferences from premises' could not be taught the simplest mathematical demonstration.

It was an odd thing, Cranmer, that you should have thought of Search's 'Light of Nature Pursued' being the book that Ellesmere was so delighted with, and scolded me for not having shown him before. This Light of Nature Pursued,'* which was written by a certain Abraham Tucker, does contain some remarkable passages bearing upon our subject. I shall give you one now, omitting certain parts of it, which, however, I hope you will be industrious enough to read afterwards. You are quite right, Cranmer, in thinking that it is the kind of book that Ellesmere would take a great fancy to.

Upon occasion of the divine care extending to the smallest things, I shall venture to put in a word on behalf of our younger brethren of the brutal species: yet it is with fear and trepidation, lest I should offend the delicacy of our imperial race, who may think it treason against their high pre-eminence and dignity, to raise a doubt of their engrossing the sole care of heaven. . . . . . . Since then, as well by God's special injunctions as by His ordinary Providence,

---

* Tucker's *Light of Nature Pursued*, vol. 5, part 3, chap. 19, edit. of 1777.

he calls upon the creatures for their labours, their sufferings, and their lives, in the progress of His great work of the Redemption, why should **we think it an impeachment** of His equity, if He assigns **them wages for all they undergo** in this im-**portant service?** Or an impeachment **of His** power and His wisdom, **if such wages accrue to them by certain stated laws** of universal nature running through **both worlds.**

In what manner **the** compensation **is operated** would be needless **and** impossible to ascertain; **perhaps they stand** only one **stage** behind **us** in the journey through matter, and as we hope to rise from sensitivo-rational creatures to **purely** rational, **so they** may **be advanced from** sensitive to sensitivo-rational. **And** when **our nature is** perfected, we may be **employed to act** as guardian **angels for** assisting them in the improvement **of** their new faculties, **becoming** lords, **and** not tyrants, of our new world, and exercising **government by** employing **our** superior skill and power for the benefit **of the** governed: by which way may be comprehended how they may have an interest of their own **in** everything relative **to** the forwarding of our Redemption. Yet **it is not** necessary **they** must have bodies shaped, limbed, **and sized exactly like ours:** for the treasures **of wisdom are not so scanty that** we should pronounce with Epicurus, there can be **no spice of** reason **or reflection, except in** a human figure, and **upon the** surface of **an earth** circumstanced just like this we inhabit.

We cut rather short our quotations from the poets on the assumption that, as men of sensibility, they would be sure to be on our side. I think, however, that when they have indulged in prose, and when they have devoted whole essays to the subject of the treatment of animals, they deserve to be quoted. Now the poet

Pope has written such an essay. It is in the 61st number of the *Guardian* of the year 1713. It is a beautiful essay, and is admirably written, affording another instance of how well poets can write in prose when they condescend to do so. He commences thus :—

I cannot think it extravagant to imagine that mankind are no less in proportion accountable for the ill-use of their dominion over creatures of the lower rank of beings, than for the exercise of tyranny over their own species. The more entirely the inferior creation is submitted to our power, the more answerable we should seem for our mismanagement of it; and the rather, as the very condition of nature renders these creatures incapable of receiving any recompense in another life for their ill-treatment in this.

He then enters very carefully into the subject, giving various instances of needless cruelty in man; and supporting his own views by reference to those of Montaigne, Locke, the Abbé Fleury, Plutarch, an Arabian author named Telliamed, Ovid, Dryden, and Pilpay. One other extract of his own writing I must give you.

There is a passage in the book of Jonas, where God declares his unwillingness to destroy Nineveh, where methinks that compassion of the Creator, which extends to the meanest rank of his creatures, is expressed with wonderful tenderness. 'Should I not spare Nineveh, that great city, wherein are more than six score thousand persons . . . and also much cattle?' And we have in Deuteronomy a precept of great good-nature of this sort, with a blessing in form, annexed to it, in those words; 'If thou shalt find a bird's nest in the way, thou shalt not take the dam with the young: But thou shalt in any wise let the dam go; that it may be well with thee, and that thou may'st prolong thy days.'

As I have referred to the essayists of former days, I must give you a passage from an essay written by the tender-hearted Steele. In the course of the essay, he objects to any representation of cruelties being produced upon the stage; and thus concludes:

The virtues of tenderness, compassion, and humanity are those by which men are distinguished from brutes, as much as by reason itself; and it would be the greatest reproach to a nation, to distinguish itself from all others by any defect in these particular virtues. For which reasons, I hope that my dear countrymen will no longer expose themselves by an effusion of blood, whether it be of theatrical heroes, cocks, or any other innocent animals, which we are not obliged to slaughter for our safety, convenience, or nourishment. When any of these ends are not served in the destruction of a living creature, I cannot but pronounce it a great piece of cruelty, if not a kind of murder.*

Mrs. Milverton. I think I may now venture, as the quotations have been less learned lately, to ask leave to read a passage from the life of my favourite author, George Herbert, written by Isaac Walton.

Ellesmere. I am not sure that I shall not take an

---

* *Tatler,* No. 134.

objection, Mrs. Milverton, to anything that has been written by Isaac Walton. I do not like that man, and I think that Byron was not too severe when he called him 'a cruel coxcomb.'

Milverton. That does not impeach the value of his evidence as a biographer.

Mrs. Milverton. Certainly not; at any rate you must let me read you the passage.

In another walk to Salisbury, he saw a poor man with a poorer horse that was fallen under his load; they were both in distress, and needed present help, which Mr. Herbert perceiving, put off his canonical coat, and helped the poor man to unload, and after to load, his horse. The poor man blessed him for it, and he blessed the poor man; and was so like the good Samaritan that he gave him money to refresh both himself and his horse; and told him, 'that if he loved himself, he should be merciful to his beast.' Thus he left the poor man, and at his coming to his musical friends at Salisbury, they began to wonder that Mr. George Herbert, who used to be so trim and clean, came into that company so soiled and discomposed, but he told them the occasion: and when one of the company told him 'he had disparaged himself by so dirty an employment,' his answer was 'that the thought of what he had done would prove music to him at midnight; and that the omission of it would have upbraided and made discord in his conscience whensoever he should pass by that place: for if I be bound to pray for all that be in distress, I am sure that I am bound, so far as it is in my power, to practise what I pray for. And though I do not wish for the like occasion every day, yet let me tell you, I would not willingly pass one day of my life without comforting a sad **soul**, or shewing mercy, and I praise God for this occasion. **And now let us** tune our instruments.'*

Milverton. **When I was quoting from the life of St. Francis, I** ought to have given **you a remarkable extract from a work which I** am told **had great authority in the middle ages.** It is 'The Revelations of St. Bridget.' This book was attacked by the celebrated Gerson, **but** was formally approved by **the Council of Basle. Here is the** passage :—

Let a man fear, above all, **me, his God, and so** much the gentler will he become towards my creatures and animals, on whom, on account of me, their **Creator,** he ought to have compassion ; for to that end was rest **ordained for** them on the Sabbath.†

Lady Ellesmere. **I wonder, Leonard, that** you have **not** given us anything **from the Spaniards, who are such** great favourites of yours.

Milverton. I could have given you many **quotations from the great** Spanish writers, including **Cervantes, for you must remember** both Don Quixote and Sancho Panza had a great regard for animals.

But, to **tell you the truth, I was a little** disgusted by my research into **Spanish authorities,** at finding that it was a Spaniard, **one Gomez Pereira, who, in** com-

---

* *Remains* **of George** *Herbert.* **Life;** p. 65.
† *Digby's Compitum.* Book ii. p. 57.

paratively modern times—that is, before the age of Descartes—had devoted a work to announcing the discovery that animals had no feeling. Indeed, it is said that Descartes borrowed his ideas from this man; but that is a fable. It was entirely consonant with Descartes' metaphysical theories that he should maintain that animals had no feeling. And what will a man not do, or say, to maintain intact his pet metaphysical theory, which is to explain all the conditions of the universe?

Sir Arthur. How one's views change as regards who is terrible, who is to be feared in this world. Time was, when I innocently thought that the description given by the great German poet, Uhland, of a wicked king, brought before my mind the most dangerous being that could well be conceived:—

> Denn was er sinnt, ist Schrecken; und was er blitzt, ist Wuth;
> Und was er spricht, ist Geißel; und was er schreibt, ist Blut.
>
> For what he thinks, is horror; and what he looks, is wrath;
> And what he speaks, is scourges; and what he writes, is blood.

But now a much more fearful creature to my mind is a 'hidebound' pedant in power (perhaps in other respects a smooth, kind, good sort of man), cursed with a few distinct ideas, whether they relate to religion, education, charity, taxation, or any other matter which greatly concerns the welfare of mankind. That is the man really to be feared—the man who is capable of doing the greatest mischief in his generation. And even if he is not in power, but has great intellectual influence, he is still a most alarming phenomenon.

Ellesmere. Especially if he is not given to admitting any exceptions. But to return to Spain. My lady did not appeal to me about Spanish authorities, though she knows that when I was a juvenile I travelled with Milverton in Spain, and that I am completely versed in Spanish literature.

Allow me to give you a Spanish song relating to animals, which I often heard my muleteers sing. It begins with the words—

> Mi muger y mi caballo—

'My wife and my horse.' It proceeds by enumerating their respective merits. It then goes on to say what a great loss it would be if he, the singer, were to lose one of these creatures, and what a small loss it would be if he were to lose the other, which could so easily be replaced. I forbear to say which is 'the one' and which is 'the other.'

Hereupon there ensued one of those conjugal quarrels of a humorous kind which have so often taken place in our presence between this well-assorted couple; and, amidst the general laughter of the company, the sitting was about to be broken up, when Sir Arthur begged to be heard.

Sir Arthur. You all know what a liking I have—almost a mania—for seeing what conclusions we have arrived at after any discussion. Of course we are all against that absurd theory of Descartes, that animals have no feeling. We will not listen to Seneca and the 'Angelic Doctor' when they maintain that animals cannot indulge in a feeling of anger which is similar to our own, or that they are incapable not merely of affection but of what we call benevolence. We are agreed upon the statements that they have keen memories (as keen, Milverton holds, as that of the late Lord Macaulay, which I take to be somewhat of an exaggeration); that many of them are highly sensitive creatures; that this sensitiveness, which is often a very disturbing force to their masters, must be dealt with by kindness and not by blows; that, according to Cranmer, the education of young people is not to be without some definite instruction as to the nature of animals, and as to the best modes of treating them. We hear, with some approval, that Sir John Ellesmere says, that animals have a lively appreciation of fun and courtesy. By the way, if they were dull creatures, I am afraid that **Sir John would not speak** quite so affectionately **of** them. **We find** that the poets, and all men of sensibility, rate animals highly, and are fond of them. We are inclined to agree with Milverton, Voltaire, **Mrs.** Jameson, **and** sundry other **persons,** who say that they do not care so much about what animals think, but that what they *suffer* is the question. Lastly, we **are** pleased to hear, without pledging ourselves to an **exact** approval of details, that Milverton puts forward certain **plans** and projects for the better treatment of animals, having in view that their owners should be better informed about their habits, **and** that these owners should have the opportunity **of** learning how their deputies and hirelings treat the animals entrusted to their care; that great heed should be taken as regards the transit of animals used for food, both on their account and our own; that the keeping of pets **should be** discouraged (against which **doctrine I for one** enter a protest); and, generally, that the observation **of the habits,** ways, and manners **of** animals would **lead to a much** more friendly relation being kept up between that **first of** animals, man, and the rest of the animated creation.

Have I not given you something like a summary **of** the case? Oh! I forgot **to mention** that Sir John Ellesmere thinks that, if there were such a thing as transmigration, **all women** would be butterflies; and that he declines to say, touching a certain Spanish song, whether it was the death of the horse, or of the wife, which would be the greater loss to the muleteer. Again I have forgotten to say that the dicta of Anaxagoras, Xenophanes, and David Hume passed with the least question from this intelligent company.

Milverton. I desire to say something more, which I especially wish to be added to your summary.

I can hardly express to you how much I feel there is to be thought of, arising from the use of the word 'dumb' as applied to animals. Dumb animals! What an immense exhortation that is to pity! It is a remarkable thing, that this word dumb should have been so largely applied to animals, for, in reality, there are very few dumb animals. But, doubtless, the word is often used to convey a larger idea than that of dumbness, namely, the want of power in animals to convey by sound to mankind what they feel, or perhaps I should rather say the want of power in men to understand the meaning of the various sounds uttered by animals.

But as regards those animals which are mostly dumb, such as the horse, which, except on rare occasions of extreme suffering, makes no sound at all, but only expresses pain by certain movements indicating pain —how tender we ought to be to them, and how observant of these movements, considering their dumbness!

The human baby guides and governs us by its cries. In fact, it will nearly rule a household by these cries; and woe would betide it, if it had not this power of making its afflictions known. It is a sad thing to reflect upon, that the animal, which has most to endure from man, is the one which has the least power of protesting by noise against any of his evil treatment.

Ellesmere. There is one thing to be said in favour of the dumbness of animals. If they could talk, it would probably be only the talk of foolish men and women. As it is, we give animals credit for an amount of good sense, which credit they might easily lose if they had the talking faculties.

Mauleverer. Considering that you profess to be such a friend to animals, Ellesmere, I do not think it is very friendly on your part to conjecture that they would talk nonsense if they could talk at all.

Ellesmere. Yes, it is wrong of me to calumniate them in this way. I'm sure I ought to be grateful to animals, for it is owing to my defence of an animal that I am tolerated here. Milverton and I were boys together at school, but I do not know that we had any particular affection for each other; and, to tell the truth, I am not sure that I am exactly the sort of person whom Milverton would have chosen as a friend. Underlying a thin upper surface of apparent toleration and readiness to hear what everyone would like to say on any subject, there is a certain real dislike to opposition in a certain quarter; and affection will not flow out to you from that quarter, if you continue resolutely to maintain that opposition.

Milverton. Most unjust! But go on with your story, for I do not know to what you are alluding.

Ellesmere. Why, the fight we had with those three town boys.

Milverton and I agreed to go out on a half-holiday with a leaping pole, to a place called Chalvey Ditch. I was to leap, and he was to look on, for he was never much given to athletic pursuits.

When we came to that renowned ditch, we found three 'town-boys,' each of them bigger than either of us, who were passing their time in the delicate and humane amusement of drowning a cat by degrees. Poor pussy had a collar on, to which was tied a piece of string. She was occasionally thrown into the water and dragged out again, for joys of this kind are not to be consummated quickly. Educate these ruffians, Cranmer! Educate them; write the book you and Sir Arthur threatened the world with.

Well, Milverton, as you might expect, began reasoning with the boys; talking to them like a Dutch uncle (I wonder what that expression means) about their cruelty.

Lady Ellesmere. **Do get on** with your story, John. Parentheses are odious.

Ellesmere. The effect produced upon the 'town boys' by **Master** Milverton's discourse was infinitesimally small. What little **effect** it had, was, I think, to increase their enjoyment. **We then** tried what money would do, and **offered the** whole of our combined fortune (one shilling and threepence **halfpenny**, if I recollect rightly) **for the** purchase of the cat. This offer was persistently refused. If it had been half-a-crown, we were given to understand, it would have been accepted. **Lady** Ellesmere will forgive me for the parenthesis; **but** I must observe that **one** and twopence halfpenny divided by three seems **to be** the precise sum which, in the boyish mind, **would** turn the balance from cruelty to avarice. Anyhow, **it was so** on that occasion.

Milverton looked expressively **at me.** That expressive **look was** fully equivalent **to** any one of the **speeches** which **Homer's heroes** make before commencing battle. He then uttered the formidable word *Nunc*; rushed at the boy nearest him, who went over into the ditch, which was pretty full **of water.** I hit the boy who **had hold of the cat as** hard a blow as I could in **the face,** and kept 'pegging away,' to use a presidential expression, with all my might. He soon dropped the cat, and did not seem inclined for further battle. Both of us then attacked the third boy, who had hitherto been a passive spectator; but he was a coward, and took to his heels at once. I clutched up the cat, Milverton the leaping pole, and off we went with our prize. After Milverton's antagonist had got out of the ditch, we were hotly pursued. The country round about is intersected with ditches, and the leaping pole was of great use to us. Home we came gloriously, and brought the cat to our Dame. It turned out to be the pet cat of a neighbouring Dame, who rewarded us with a bottle of currant wine and twelve twopenny gooseberry tarts.

I have not indulged much in currant wine since, but gooseberry tarts have a wonderful attraction for me to this day; and when I do but see a gooseberry tart, the glories of the past come back upon my soul, I scent battle from afar, and feel like one of Ossian's heroes.

Milverton took to me amazingly ever afterwards, was always ready to give me two or three sentences for the weekly Latin theme; and here I am, snugly ensconced in this house, owing my not altogether unfavourable reception chiefly, I believe, to my ready backing of my friend in this encounter with the town boys. I must own that the ditch was greatly in our favour, for Milverton was never good at fisticuff work, and we should have been ignominiously thrashed but for his having the bright thought of making his antagonist—not bite the dust of the plain—but imbibe the muddy waters of the Chalvey.

CHAPTER VI.

After the conversation given in the preceding Chapter, we all went out for a row upon the neighbouring lake, which Sir John Ellesmere, to the great annoyance of Mr. Milverton, would always call 'the pond,' although it consists of nine acres of water, and was even put down in the county maps as 'a lake.'

While we were upon the lake, conversation was resumed. I could not well take note of all that was said, being employed at first as one of the rowers; but when I was permitted to be a sitter, I succeeded in making some notes, the most important of which were as follows:—

Mauleverer. The last vessel I was in, was of a somewhat different description from this. A few weeks ago I went over one of our first-rate ironclads at Portsmouth. I had not been in any great vessel for many years; indeed, my previous experience of warlike craft was that which I gained in going over the 'Duke of Wellington,' when she was first commissioned.

If you have not seen any of our more recent men-of-war, you can hardly imagine the extraordinary strides made in scientific inventions applicable to naval warfare. Everywhere throughout the vessel, even in the most trifling matters, there is amazing ingenuity shown. I felt almost suffocated with contempt for myself and my species, when I contemplated this ingenuity. The vast absurdity of mankind rose up before me in a more visible and trenchant manner than it had ever done before.

Ellesmere. **What an** anti-climax ! I really thought that Mauleverer, after admiring something, was proceeding, however painful the process might be to him, to admire somebody.

Mauleverer. I thought of the squalid streets and miserable alleys in which so large a portion of our town population dwells. I thought of the mean hovels in which many of our peasantry abide. I thought—it may be a commonplace way of thinking—of how clever we are in slaughter, and how stupid in self-preservation, and in maintaining the real dignity of man by beautiful modes of living. My contempt was universal. It involved all Europe—the most civilized portion of the globe, as we fondly call it—in its comprehensiveness; and, finally, I thought of Schiller's grand saying, 'The world's history is the condemnation of the world.' *

Milverton. I should fully sympathise with you if you would strike out the word 'contempt,' and use some such word as sorrow, instead.

Sir Arthur. What horrifies me is, that certain proceedings, which have taken place of late years, make our future prospects of peace so doubtful. I do not believe that a greater error in high statesmanship could well have been committed than by imposing huge pecuniary fines upon conquered nations.

Cranmer. I can't agree with you. These fines are restraints.

Sir Arthur. That, my dear Cranmer, is, I fear, but a shallow view of the case, and one which, to the best of my belief, history does not bear out. Nations are not prevented from going to war by poverty.

Milverton. I cannot call those men great statesmen who adopt such remedies for keeping peace. I picture to myself what the greatest minds who have given themselves to statesmanship would have said of such transactions. How foreign they would have been to the plans of a Bacon or a Machiavelli.

Ellesmere. Yes; the virtuous Machiavelli, whom

---

* Die Weltgeschichte ist das Weltgericht.

Milverton so much admires, would have said, 'Slay as many of your enemies as you like, but do not impoverish them. Slaughter is soon forgotten: pecuniary mulcts are neither forgotten nor forgiven.'

Milverton. It is very unfair to put such a saying into the mouth of Machiavelli; but that he would have objected to pecuniary mulcts, I am sure.

Ellesmere. Ignorant as I am both of Bacon and Machiavelli, I shall put the matter in a more homely way, but one that I deem will not be less convincing. Imagine a fight between two schoolboys. If the victorious schoolboy, after thrashing his antagonist, were to take away his little pocket-money and his knife, do you think those two boys would ever be friends again? Whereas, the surest way to establish a firm friendship between two boys is, that they should have a good fair stand-up fight. I wish I had had a fight with Milverton when we were boys; we should be much more affectionate friends than we are now!

Sir Arthur. I think that Ellesmere's simile is excellent.

Milverton. And I must say for Bacon and Machiavelli, that their peculiarity is, that they always take the wisdom that lies at their feet, and they would not have despised even Ellesmere's boyish simile. The difference between these wise men and other men is just this, that they seize upon what is common and self-evident, and make it almost uncommon and most searching. Now I will give you a simile that occurs to me in reference to Mauleverer's ironclad. It is that an ironclad and a cathedral very much resemble one another.

Ellesmere. That *is* an astounding simile.

Milverton. You must look a little beyond the immediate objects, before you recognize the truth of it. Each of these wonderful productions is the outcome of an age which is mainly directed in its thoughts to the products in question. You could not have had those cathedrals, if almost the whole mind of the men of that time had not gone in the direction which compelled those cathedrals to be built. You could not have these ironclads, and all this wondrous skill devoted to slaughter, which characterises this age, and moves the contempt of Mauleverer to such a height, if self-defence had not become one of the most urgent, perhaps *the* most urgent, necessity for nations.

Sir Arthur. And what becomes of Christianity?

Mauleverer. What, indeed!

Milverton. I wish to go back for a minute or two in our conversation. I want to drive thoroughly out of Ellesmere's head the notion that these great writers, of whom he is quite ready to speak disrespectfully, were merely subtle thinkers, or that their thoughts are, for the most part, far-fetched. I do not believe that mere subtlety in a writer has ever commanded the world's esteem.

Sir Arthur. No; when Bacon says 'reading maketh a full man; conference a ready man; and writing an exact man,' there is no particular subtlety in what he says, and you are ready to give assent to the saying the moment you have heard it.

Milverton. I remember a passage in Machiavelli respecting which the same assertion, as it seems to me, may justly be made. He is discussing the dangers to a prince from conspiracies, and he says that the danger arising from private injuries is immensely increased if the prince is generally unpopular. Therefore, he says, let a prince take care to avoid this general unpopularity. If he succeeds in this, he will run much less risk from the effects of private injuries, whether they be such as have touched estate, life, or honour. The private injury remains the same; but even if there should be the desire and the power to revenge it by means of conspiracy, the conspirators are restrained by that universal favour which they perceive attaches to the prince.*

Sir Arthur. Yes: how far from subtle is this

---

* Che quando pur ei fussero d'animo e di potenza da farlo, sono ritenuti da quella benevolenza universale, che veggono avere ad un principe.—*Discorsi,* lib. 3, cap. 4.

remark. We must admit the truth of it on the most ordinary occasions in life. You will find that you are much more prone to resent any injury which has been inflicted upon you by an unpopular person. The general disfavour often seems to increase the injury, or, at any rate, there is not the restraint upon the injured man which he is sure to feel if he receives but little sympathy from others by reason of their general approval of the injurer.

Milverton. I wish to carry the subject much further. The sayings of these great men are generally well-clothed in words; but I wish to exalt the merit and force of sayings which have not this advantage.

I think that some of the most fruitful, if not the wisest, things that have ever been said, are of this nature. They are not exactly commonplace things; indeed, when first uttered, they may have been the product of severe thought; but they are, for the most part, statements of facts, statements which cannot be denied, which do not strike you as very wonderful when you first hear them, but which grow in importance as you think over them, and as they are illustrated by the facts that daily come before you They are not sharp sayings, which generally have some narrowness in them, but they are sayings of an immensely broad character. There is even a certain bluntness about them.

Ellesmere. This is **all very fine**, I daresay; but I have not the most dim **notion of what** the man means.

Milverton. I must take **an instance.** It has been said that one of the best means to **ensure** contentment, is to take care that you do not set your heart upon something which must, of necessity, **be** given you by other people. You will doubtless find **scores** of writers, **from** Epictetus downwards, who have uttered this saying in various forms of language.

I must proceed with further illustration, for I fear that there is still some haziness about my meaning. Suppose a man, for example, has made the main pleasure of his **life to** consist in applause. He is completely at the mercy of his fellow men. He may do something ever so well; but, 'contrive' them (I will explain that word afterwards), **they** will not clap their hands. And as this applause of other people is what he **has** set his heart upon, and not the doing **of something** to his own satisfaction, he is entirely dependent upon the **will, or** perhaps the wilfulness, **of his fellow** men.

I do not think that you can exaggerate the fruitfulness of this thought, only it requires **to** be much dwelt upon, in order to **bring out its full merit.** Of course, it has nothing to do with bread-winning pursuits. If you **are to** get your living by making knives or pins, or giving legal opinions, you have to produce knives, pins, and legal opinions, which shall satisfy your fellow creatures, and ensure a good sale for these articles. But if, over and above this necessary repute, you hunger for praise, you have certainly given an additional hold upon yourself to your fellow creatures, which, as I said before, places you entirely at their mercy.

Take authorship, or statesmanship, or any of the more refined modes of labour, and this same rule applies. That man has put himself so far into the power of other people, who beyond caring for doing his work to the best of his power, looks to popular applause for his reward.

You will find that a number of wise and witty sayings are all contained in, and indeed anticipated by, this one thought, illustrating the danger of making your contentment depend upon what others alone can give.

Sir Arthur.

'Tis not in mortals to command success,
But we'll do more, Sempronius, we'll deserve it.

—only one ought to substitute the word 'applause' for 'success.'

Milverton. Yes, Sir Arthur; but the number of such sayings anticipated by this original maxim is legion. It chiefly applies to that which is, if I may so express it, extra to the ordinary duties and works of life—to that which gives them their especial savour to the man who does them. If this savour mainly depends upon recognition, and praiseful recognition, he is to that extent a slave. We ought to be able to say, in Churchill's admirable words—

> 'Tis not the babbling of an idle world,
> Where praise and censure are at random hurled,
> That can the meanest of my thoughts control,
> Or shake one settled purpose of my soul.

Sir Arthur. I am thinking of Gibbon. He was a man who did not surrender his contentment into other men's keeping. Early in life he felt that he was a writer whose gifts by nature could best be made use of by writing a history. He looked about for a subject which would interest, or ought to interest, the world. As I dare say you know, he did not choose 'The Decline and Fall' at first; but when he had chosen it, he abided by it; and I feel confident that his contentment depended upon working out his own purpose —that he was thenceforth no man's slave, that he was dependent upon no man as regards censure or applause.

Ellesmere. I now want to know why Milverton used that strange word 'contrive' so inappropriately. I suppose there is some pedantic reason for it.

Milverton. I will justify my use of the word. There was a senior Fellow, holding high office in our college—one of the best of men. He would tutorise a poor Sizar without receiving any payment, if he saw any worth in the young man. He was of a warm temper, and very much given to the love of discipline and decorum. His notions of decorum were sometimes outraged by us, as, for instance, in our attendance at morning chapel. He would come into my room, or that of some man he knew, and lecture us somewhat after the following fashion :—' I don't say that you are late for chapel, but you don't take your proper places at once; you hang about the doors, loiter, and gossip; and, con-con-con-contrive you, you keep the Master and Senior Fellows waiting in the ante-chapel.' The good man longed to say the word ' confound,' but thought it would be indecorous; and the nearest substitute which occurred to him, when he was angry, was ' contrive.' I have ever since adopted the word as a delicate and decorous substitute for what might otherwise be considered as swearing.

Ellesmere. I should have thought that a calm-minded philosophic man would never even have thought of using the word 'confound.' For my part I always say something much stronger or much feebler. It is not a favourite word of mine.

But let that pass. I want to ask a pregnant and personal question. When you write anything, Milverton, what are your feelings anent criticism, particularly as regards hostile criticism? To put it plainly: if there is a man in the world who desires to persuade and influence other people, you are the man! Consequently you have placed your contentment in other people's keeping, and you are, to use your own polite expression, 'a slave.'

Milverton. Not at all. You force me to speak egotistically; but I can give a perfect answer to your question.

When I write anything, it is with a purpose. I own that I hate to have my purpose thwarted. If a critic, commenting upon what I write, agrees with me, I honestly admit I am pleased. He has furthered my purpose.

Et sapit, et mecum facit, et Jove judicat æquo.

I am very much obliged to him, and am disposed to like him.

If he opposes me, and there is anything good and serviceable in what he says, I take note of it for future occasion. If he is only objectative and Ellesmerian, I say to myself, 'Contrive the fellow.' I regret he is not with me; indeed, if you like to put it so strongly, I am vexed at this hindrance to my purpose. But I have not placed my contentment in the measure of applause one may get for anything.

Do any of you understand the system of 'double entry?'

Cranmer. I do.

Ellesmere. The rest of us, by our silence, show that we do know it, but that we are not, like **Cranmer**, anxious to parade our knowledge of it.

Milverton. It is but too evident that Ellesmere knows nothing at all about it. I will explain. Correct me, Cranmer, if I do not do so rightly.

You must suppose yourself to be a merchant, and to be able, by this excellent system of double entry, to appreciate, if not to ascertain, at any moment, the result of every venture separately. The good ship 'Mary Ann' has gone to Santander with a cargo suitable for the Spanish market. The good ship is treated in the merchant's books as a human being would be. She has her 'debit' and her 'credit;' and this mode of keeping accounts shows whether the venture on the 'Mary Ann' has been successful or unsuccessful.

Now let a man who works in politics, in literature, or in art, deal with his Bill, or his book, or his picture or statue, as the merchant does with the good ship 'Mary Ann.' Let him put down everything that is to its debit or its credit; but at the same time take care to keep it as a separate transaction, in which the whole of his fortune is not involved. He will find this mode of looking at his liabilities and his ventures to be most useful and comforting. The good ship does not always come to port, or she does not always find a ready market for her cargo, however well chosen. Other people's conduct, or their prejudices, just or unjust, may interfere with success; but if he has adopted the practice of keeping the accounts of this transaction separately, he knows where he is in the venture, and seldom allows himself to consider it, however deplorable the result, as a wreck of his fortunes.

In political action there is more comfort to be had from that astute thinker, De Quincey, than from almost any other man. He looked upon the transactions in politics as the resultants of certain forces. A minister bringing in a Bill, should do the same. The forces may be such as neither wit nor wisdom can overcome. On the other hand, they may be such as to drive on his Bill to a successful issue. In no case should he suffer himself to be utterly disheartened by hostile criticism, if he has done his best to promote his object. The same with the sculptor, the painter, or the author. He must not fix his vanity upon the thing attempted, only his intention and his purposefulness (if I may coin such a word); and you will find that most people will bear the thwarting of their intentions and their purposes better than the lowering of their vanity and their self-love.

Ellesmere. What an admirable consoler Milverton would make for unsuccessful people.

Your mercantile mode of putting the thing is very well in its way, but I could say something much more to the point. There is a little anecdote which I have heard before in this company, but I shall refresh your memories by telling it again.

There was a certain great man, the head of a house, with numerous scions belonging to it. He was a very wise old man. I used to notice, when I was a boy, that he was one of the few persons whom the late Duke of Wellington would take a walk with. You will naturally think, how should I know he was a wise man?

Lady Ellesmere. Do get on with your story, John.

Ellesmere. What an impatient little woman you are! I wonder from whom you learnt the habit of impatience.

I was employed in a great case. My junior begged me to read this man's evidence, which had been given before a Parliamentary Committee, as it bore upon the case. The evidence was masterly. In general, we lawyers are not only great in cramming for a case (we should be the boys for competitive examinations), but we are also equally great in discumbering our minds of what we have crammed up for the occasion. But this man's evidence remains to this day in my mind.

Well, there came a general election, and several of the scions of that house lost their elections. One, however, not the wisest of the family, succeeded. We will call him Leonard. 'How is it,' said one of the chiefs of the party to this old man, 'that John, and Robert, and Thomas, failed, but Leonard succeeded?' 'Oh! his nonsense suited their nonsense,' was the apt reply. And how much of human life it explains. You don't win people by talking their sense to them, but by talking their nonsense to them, which they are fondest of. And then, if you can talk the right nonsense to the right people, and at the right time, you must succeed. Sometimes one is aghast with astonishment on reading a speech of vast and continuous folly, which one finds met with great applause. The truth is, it does not happen to be one's own nonsense, but it is theirs— the people whom the orator is addressing.

Cranmer. This is all very witty and very satirical, I daresay, but what consolation there is in it, I do not exactly perceive.

Ellesmere. Why, don't you see that the unsuccessful man can always comfort himself by thinking that his sense did not suit 'their nonsense'? You may be sure that there are thousands of people who console themselves, and sometimes justly, by saying to themselves that they are before] their age, or, at least, that they are out of joint with their age. I have scarcely a doubt that Mauleverer, with whom we hardly ever agree, retains his placidity while amongst us, by thinking that his nonsense does not suit our nonsense, or, as he would disrespectfully put it, that his sense is prematurely confided to us. Just the same thing happens in art and in literature as in talkee-talkee. Only the word 'taste' must be substituted for 'sense.'

While we were in the boat, evening came on, and with it one of those grand sunsets which are not seen many times in a man's life, and which remain strongly impressed upon his memory, although he have but little power of describing them to other persons.

This was no mean sunset occupying only the western sky, but it overspread the whole range of the heavens. A new world seemed to be developed in the skies, a world wherein there were vast flame-coloured mountains, piled up, one over the other, with wide still seas of that peculiar yellow-green colour, so rarely to be seen, and on which large dark islands seem to float. Every shade of colour and every kind of form were there. Dimly in the east a soft purple haze was visible, while in the west there was a livid redness. The full moon, as she slowly rose from the horizon, gathered to herself all that was softest and most beautiful in the colouring of the heavens, and formed the most exquisite contrast to her fierce brother, the sun, who left the world in solemn anger.

Silence came not only upon the 'Friends,' but upon all animated creatures near us. Nothing was heard but a soft rippling sound among the reeds, where, at the moment, we had stayed the boat, and the occasional fall of drops of water from the suspended oars.

Then, almost suddenly, the resplendent colouring grew dim, and all that remained in the western heavens were dark masses of cloud, and in the east the now brightly-shining moon.

The conversation was resumed, but the sunset had produced its effect; Sir John Ellesmere remained silent. There was an occasional wail in words from Mauleverer, which was disregarded, and the talk was left to Sir Arthur and Mr. Milverton. I cannot distinguish the parts they severally took in it, but the result was something of this kind :

They spoke of the Cosmos of Humboldt; of

the whole order of nature; of its unity; of what it might teach us if we could but ascertain its few great laws, and abide by them. They said how scraps of morality and shreds of doctrine would be absorbed and rendered needless, if we could but once appreciate the meaning, the order, and the beauty of the universe, as a whole. They reverted to the subject we had been so much considering, namely, the treatment of animals, and said how little occasion there would be for maxims and rules and laws about it, if only we had learnt the real relationship that existed amongst all animated beings.

They spoke of the views of a great German writer, who had said that what is highest and noblest in man, conceals itself, and is without use for active life (as the highest mountains bear no herbage), and out of the chain of beautiful thoughts only some links can be detached as actions.*

But they contended that even this was but a

---

* Das Höchste und Edelste im Menschen verbirgt sich und ist ohne Nutzen für die thätige Welt (wie die höchsten Berge keine Gewächse tragen) und aus der Kette schöner Gedanken können sich nur einige Glieder als Thaten ablösen.—Jean Paul Richter, *Hesperus*.

contracted view, or at least one that required further development; for, that these high moods of thought, which, above all things, the appreciation of nature in its most beautiful forms tended to produce, would render the application of mere rules perfectly unnecessary, as the spirit in which men would act, would overcome and be beyond all written rules and regulations.

So far my mind went with them, but could not quite follow them, when they went into what appeared to me to be wild hopes and aspirations for the future; imagining a time when all nature would be better understood, and men be drawn more near to Nature's God; when most of the tribulations arising from wars and contentions, and the clashing jealousies of sects and classes amongst mankind, would be impossible by reason of their then transparent folly. I could not partake of this hopefulness, which belongs, I suppose, to the poetic temperament. Would that prophetic power did but accompany poetical temperament!

The stars came out vividly in those parts of the heavens where the clouds were not massed together. Mauleverer, who has studied astronomy, reminded us of some of the great discoveries connected with that science, oppressing us rather with descriptions of the magnitude of the universe, and of our own smallness and contemptibility. This kind of discourse was, no doubt, very gratifying to him, but did not tend to impart cheerfulness to the rest of us. And then we rowed to the landing-place, and walked up in silence to the house.

CHAPTER VII.

It was rather surprising to me that our friends had hitherto, with the exception of the conversation in the boat on the lake, kept even so closely as they had to the subject which Mr. Milverton put before them at the beginning of the holidays. I knew that it was pain and grief to Sir John to keep so closely to any subject. We had, however, another morning devoted to it, and the conversation chiefly turned upon Eastern thought and action as regards animals. Sir William Jones, Professor Max Müller, Professor Wilson, and other great authorities in Eastern lore, were, together with the Koran, very largely quoted; but I really think we have had so much of learned quotation brought forward, that my readers will not care to have this conversation reported to them verbatim. The only quotations I shall give are those which Sir Arthur made from the Koran and its notes, and which were the following :—

There is no *kind of* beast on earth, nor fowl which flieth with its wings, but *the same is* a people like unto you; we have not omitted anything in the Book *of our decrees*: then unto their Lords shall they return.

Whereupon Sale, the translator, in a note applying to this passage, says :—

For, according to the Mohammedan belief, the irrational animals will also be restored to life at the resurrection, that they may be brought to judgment, and have vengeance taken on them for the injuries they did one another while in this world.

Sir John Ellesmere did not fail to remark that it does not seem to have occurred to the translator that any vengeance was to be taken for the injuries which the other animals had received from the hands of man.

Our own Scriptures were quoted, and Mr. Milverton insisted much upon the passage :—

Are not five sparrows sold for two farthings, and not one of them is forgotten before God?

Sir John Ellesmere then made an attack on Mr. Milverton, saying that he had not given enough attention to providing practical remedies for the evils he had enumerated as regards the treatment of animals. I was sorry to hear that Mr. Cranmer joined in this attack. Mr. Milverton, in reply, said something of this kind :—' Hundreds of thousands of transactions will take place in our metropolis to-day in which the treatment of animals by men is concerned; and do you think that I or anyone else can lay down rules by which these transactions can be regulated? If we could lay down such rules, would they be obeyed? Our object should be to get a better spirit introduced into the treatment of animals. That is the only thing which will really have an abiding effect.

'It is very ungrateful of you, Cranmer, joining with Ellesmere on this occasion; for I admit, and I think admitted at the time, that your suggestion that a book should be written for schools, which should inculcate humanity to animals, was a most excellent one. I even said that you and Sir Arthur might write it: though I think it had better be entrusted to a Huxley, a Wallace, a Frank Buckland, a Hooker, a Wood; or to Bates, who wrote that book upon the Amazons, or some man of that kind, practised in the observation of nature.'

'You ask for practical remedies, for things to do. Well, I say, follow the example of Lady Burdett Coutts, who has given prizes for the encouragement of humanity to animals. I do not mean that you are to do precisely the same thing—to follow exactly in her steps; but read up the subject (a great deal of evidence has been given upon it before Committees of the House of Commons), and you will soon find out something to do. Meanwhile, interest yourself in the doings of that excellent 'Society for the Prevention of Cruelty to Animals,' and of other societies that have like objects in view.'

'You see that improvement in the treatment of animals depends upon many small things which it would be almost impossible to enumerate, and the value of which would only be appreciated by those who are conversant with the particular branch of the subject to which these small remedies refer. I will give you an instance. You would hardly believe, unless you had heard the evidence of experts, how much can be done to improve the transit of animals by sea by such regulations as the following, the adoption of which is recommended by the Transit of Animals Committee. "The floors of each pen should be provided with battens or other footholds; and ashes, sand, sawdust, or other suitable substance should be so strewed on the floors of the pens, and on the decks and gangways, as to prevent the animals from slipping." A whole host of evils could be avoided by these simple regulations.'

I have now given some notion of the serious part of the conversation. There was only one part of it that may be called amusing. Sir Arthur, who is quite as zealous an advocate for the good treatment of animals as my chief is, gave us an anecdote which was very characteristic. When he was in office he had to attend a function out of town, at which one of our most distinguished statesmen was also to be present. They were both great lovers of animals, and each had imparted to the other his determination never to urge the driver of a hired horse to go faster than the driver chose to go. Unfortunately, on this occasion, they were detained by urgent business in Downing Street, and started together very late in the same cab to the station. In the minds of both of them there was a terrible fear that they might be too late, and, if so, that they would keep great personages waiting, would incur much blame, and would, perhaps, cause a breaking down of the function altogether. After they had started, they pretended to talk with interest of things which were really indifferent to them; for the thought which galled both of them was whether they should be late. Furtively they looked at their watches at almost every other minute. At length they could not keep up their sham talk any longer, and there was dead silence. After this had lasted about seven minutes, one of them had the boldness to state the question which was in both their minds. The cab-driver meanwhile was taking it very coolly. They imparted to each other their hopes and fears. The best that could happen to them if they should be late, was, that they would be able to get a special train. Even that might not be feasible. Should they keep to their principle of action, or should they break it? They resolved to keep to it. The event was not disastrous. They were just in time, with not a second to spare, and they had the satisfaction of having abided by their principle. As they discussed the matter with one another, they had said, 'This appears very important to us, that we should be in time : with thousands of other people the same argument might hold good. How can we ever have the face to urge upon them that they should adopt the principle, which we break through when a certain amount of pressure is brought to bear upon us?' Sir Arthur finally said, 'That he did not know that in the whole course of his life he had undergone so severe a temptation to do anything which, upon general grounds, he thought to be wrong.' We praised him very much, but he would not accept our praise, saying, that if there had been the slightest weakness in his companion, he might have given way too, and would have insisted upon its being a justifiable exception. The malicious Ellesmere ventured to suppose that if each of these distinguished persons had been alone he would not have hesitated to urge the cabman to urge his horse. But this mean suggestion was universally scouted, and Sir Arthur became a little angry at its having been made.

Sir John Ellesmere then, somewhat to our astonishment, began to dilate upon the pleasures of companionship, and to say what a grand thing it is to be able to discuss any subject with frequent interruptions, such as could not be allowed when listening to a sermon or a speech; and then he urged Mr. Milverton to write an essay on the joys of high companionship, in the course of reading which he, Sir John, was to be permitted to interrupt at least four times. The other guests seconded Sir John's request. Mr. Milverton consented, and, when he had prepared his essay, read it one morning when we were all assembled in the study.

ON THE JOYS OF HIGH COMPANIONSHIP.

I wish, my good friends, that you had not given me so fine a title for this essay. I shall be obliged to consider many forms of companionship which cannot be described by so grand an epithet as 'high.'

The joys, not merely of high companionship, but of any companionship that is tolerably pleasant, are so great, that a man with whom all other things go ill, cannot be classed as an unhappy man, if he has throughout his life much of this pleasant companionship.

The desire for companionship is absolutely universal. Even misanthropy is but the desire for companionship, turned sour. This desire extends throughout creation. It is very noticeable in domestic animals; and could we fathom the causes of their sociability, we should probably have arrived at a solution of several important questions relating to them and to ourselves.

The most fascinating people in the world have, I believe, been simply good companions. Shakespeare, as he knew most things, knew this, and has shown that he knew it, in what he has indicated to us of the loves of Brutus and Portia, of Antony and Cleopatra, and of Rosalind and Orlando. We now know that that 'brown Egyptian beauty,' of whom Tennyson says, in his 'Dream of Fair Women'—

> When she made pause I knew not for delight;
> Because with sudden motion from the ground
> She raised her piercing orbs, and filled with light
> The interval of sound.

was no beauty at all, and indeed by many people would have been considered plain; but no doubt she was an exquisite companion. The same may be said of many of the most fascinating people, men and women, of our own country, and of our own time. However this may be, I think it must be admitted that one of the main objects of life is good companionship. "What," says Emerson, "is the end of all this apparatus of living—what but to get a number of persons who shall be happy in each other's society, and be seated at the same table?"

[*Ellesmere.* I shall take leave to make a remark upon that afterwards.]

The first thing for companionship is, that there should be a good relation between the persons who are to become good companions to each other. It is not well to use a foreign phrase if it can be avoided, but there are foreign phrases which are supremely significant, and utterly untranslatable. I therefore say that those people I have spoken of should be *en rapport* with one another. This *rapport* may have its existence in various ways. The relationship of mother and son, of father and daughter, will give it; the love that some people have for children will give it with children; similar bringing up at school or at college may give it; similarity of present pursuits may give it. But before all and above all, that incomprehensible, unfathomable thing called personal liking—that which you feel (or the contrary of which you feel), frequently at first sight — will be sure to give it. We use the phrase 'falling in love :' we might perhaps use the phrase 'falling in liking' to describe a similar unavoidable precipitancy.

The cat that purrs at your approach establishes herself *en rapport* with you; and there is a human purring, sometimes quite inaudible to alien ears, which also does not fail to establish the requisite relation.

There may, however, be very good and sound companionship without friendship, as there may be friendship of a most deep and sincere kind with but a sorry accompaniment in the way of companionship. The same may be said of lower degrees of regard than that which is expressed by friendship; and you may have but little respect or liking for one to whom you cannot deny the merit of being a good companion. There is an affinity between friendship and companionship, as there is between the metal that is moulded into a medal and that which is turned into a current coin; but the uses of the two things are very different. In cases where there is this good companionship, without a tie either of affection or regard, there is always some other tie to be found, which may be included under the head of necessary social intercourse, and serves to form the basis of the companionship.

The beginning once made, the basis once laid for this companionship, what are the qualities which tend to make it continuously pleasant?

The first thing is confidence. Now, in using the word confidence, it is not meant to imply that there is an absolute necessity for much confidingness in small things. Wilhelm Von Humboldt has expressed an opinion which is worth noting in reference to this subject. "Friendship and love," he says, "require the deepest and most genuine confidence, but lofty souls do not require the trivial confidences of familiarity."*

The kind of confidence that Humboldt means, and which is required for companionship as well as for friendship and love, puts aside all querulous questions as to whether the companions like one another as much to-day as they did yesterday.

---

* Freundschaft und Liebe bedürfen des Vertrauens, des tiefsten und eigentlichsten, aber bei großartigen Seelen nie die Vertraulichkeiten. Dr. Ramage's *Beautiful Thoughts from German Authors.* Liverpool, 1868.

Steadfastness is to be assumed. And, also a certain unchangeableness. "He is a wonderfully agreeable person," said a neighbour of one of the best talkers of the day; "but I have to renew my acquaintance with him every morning." That good talker cannot be held to be a good companion in a high sense of the word. Again, this steadfastness makes allowance for all variations of humour, temperament, and fortune. It prevents one companion from attributing any change that there may be in the other, of manner, of bearing, or of vivacity, to a change in the real relation between the companions. He does not make any of these things personal towards himself. Silence is not supposed to be offence. Hence there is no occasion to make talk, a thing which is fatal to companionship. One reason why some of us enjoy so much the society of animals, is because we need not talk to them if we do not like. And, indeed, with a thoroughly good human companion, you ought to be able to feel as if you were quite alone.

The hindrances to sound and pleasant companionship are far more of a moral than of an intellectual kind. Diversity of tastes may be no hindrance, and, on the other hand, they may be the greatest hindrance to good companionship. I think, however, it may be observed that those tastes only are fatal which have their origin in what is moral or emotional, and not in what is intellectual. For instance, there shall be two persons, excellently fitted for companionship; but very different, intellectually speaking, in their tastes. One shall be a great lover of art, and everything that is inartistic is very repulsive to him. He will comment upon what offends his taste in works of art, in literature, in all the apparatus of living. This will not hinder even the perfection of companionship, although the other is a plain blunt person, somewhat inclined to think this frequent reference to art rather tedious. If, however, he has a taste for ridicule, and indulges it in commenting sarcastically upon that which is a very deep part of his friend's nature, those two friends will soon cease to be good companions to one another. There can be no healthy companionship when there is a perpetual expectation of attack.

[*Ellesmere.* Oh !]

Unless, indeed, it is thoroughly understood that the attacks are purely playful, and that when one friend differs from the other, and manifests this difference by anything like satire, there is at the same time a confidence in the mind of the person attacked, that his friend really respects his views, and would wish to adopt them if he could.

[*Ellesmere*. Oh! indeed.]

Difference of temperament is no hindrance whatever to companionship. Indeed, the world has generally recognized that fact. We all know that the ardent and the timid, the hopeful and the despondent, the eager and the apathetic, get on very well together. What may not always have been as clearly perceived is, that there are certain diversities of nature, chiefly relating to habits, which produce, not agreeable contrasts, but downright fatal discords. And, in such cases, companionship of a high kind is hopeless.

Benefits neither make nor hinder true companionship. I may also observe that relationship ought to ensure good companionship, because, however unpleasant it may be to relations to be told that they are wont to resemble one another very much, they often do so resemble one another. And, frequently, this resemblance exists in those very qualities that tend to ensure good companionship. Why, however, relatives often fail in becoming pleasant companions to one another is, from a cause which also destroys so much love and so much friendship in the world. It is an unreasonable expectation of liking, love, or sympathy, grounded merely upon some external connection, if it may be so called, of the one person to the other. You must not expect that people will change their natures in their dealings with you, merely because they are your friends, or your relatives, or your lovers. They are thus brought into relation with you, but not necessarily *en rapport*. And, in fact, you must take more pains with them than with other people to create this *rapport*, and, if possible, to satisfy the unreason of their expectations.

One might deal with companionship under several heads—moral, intellectual, even physical. But the truth is that it is a thing 'compounded of so many simples,' that it would not be judicious, and certainly would be pedantic, to consider the subject after the fashion of the schoolmen in a distributive manner. I have endeavoured to show that good companionship depends upon many things which may be called of a minor kind, and have not disdained to attribute great weight to such a fanciful thing as personal liking.

Now let us suppose that the principal requisites for companionship have been attained; first, the basis for it created by personal liking, early association, similarity of pursuits, and the like; secondly, the means of continuing it, such as this confidence that has been spoken of, the absence of contravening tastes, the absence of unreasonable expectation, and the like. Now, for what remain to be considered as the essential requisites for high companionship, we must enter into what is almost purely intellectual. For this high companionship there must be an interest in many things, at least on one side, and on the other, a great power of receptivity. It is almost impossible to exaggerate the needfulness of these elements. Look at results. Consider the nature of those men and women whom you have found, if I may use the phrase, to be splendid companions. It is not exactly their knowledge that has made them so; it is their almost universal interest in everything that comes before them. This quality will make even ignorant people extremely good companions to the most instructed persons. It is not, however, the relation of tutor to pupil that is contemplated here. That is certainly not the highest form of companionship. The kind of ignorant person that I mean, if he or she should be one of the companions, is to be intensely receptive and appreciative, and his or her remarks are very dear and very pleasant to the most instructed person. Is not the most valuable part of all knowledge very explicable, and do you not find that you can make your best thoughts intelligible, if you have any clearness of expression, to persons not exactly of your own order, if you will only take the pains to do so?

[*Ellesmere*. I never can make Cranmer understand my best points.]

The most consummate companions, intellectually speaking, are those who do not talk much about the past. The present and the future are their main subjects. This was to be observed in the late Lord Palmerston. In addition to this quality for a good companion, he was a man who had the most intense interest in every branch of human effort. The fulness of vitality that was in him, even when he was in his 79th year, made him care for what was present, and what was to come, in all affairs of human interest, and especially in discoveries and inventions.

What constitutes a bore? Three qualifications are requisite to make a perfect bore. He must prefer hearing himself talk, to the pleasure of eliciting good conversation. The limitation of his interest in human affairs is very restricted; therefore he repeats himself largely; and, as you will observe, he is very fond of talking of the past, and of the past in the strain of Æneas, often introducing the sentiment, if not the actual words, *quorum pars magna fui.*

The reason why great literary men and great statesmen are such interesting companions, is, not only that they have a very wide range of subjects, but that they are generally very anxious to persuade their companions of something, to bring them round to their way of thinking; and their earnestness is contagious, and creates earnestness, and therefore pleasure, in the soul of the hearer. And in general, these men are delighted with any companion who is appreciative and receptive, whatever his other gifts or failings may be.

Lastly, it is to be observed, as a general rule, that almost every very great man is an adept in companionship, and is a companion to almost everybody with whom he enters into social intercourse. What a memorable description that is of Burke's excellence in that aspect—that if you had met him, taking shelter under an archway, you would have found out that you were in the company of a great man! You would, however, not only have found out that; but you would have discovered that you had met with a good companion—with one whose society you would long for, as it would fulfil all the conditions for evoking and maintaining the rare felicity of high companionship.

CHAPTER VIII.

Milverton. I think I heard you interrupt once or twice, Ellesmere. You can now make your objections.

Ellesmere. There was one part of your essay, Milverton, that was weak—decidedly weak. You were dilating upon a part of the subject which you do not understand. It was where you spoke of bores. I maintain that no man deserves the title of a bore, who does not deal largely, I may say offensively, in details. A bore over-explains, over-illustrates, molests one with needless dates, informs one of Smith, of whom he is going to tell a good story, that he is first-cousin to Jones of the Audit Office, and thinks you must have met him at Robinson's, who married into Brown's family—not the Highgate Browns, but the Browns who have that nice place at Hackney, with green gates, and a porter's lodge built after a Chinese pattern. At last you hardly know where you are; and there is a feeling of being suffocated by a profusion of facts, every one of which might, judiciously, be omitted. 'Le secret d'être ennuyeux c'est de tout dire.'

Why it is that I say you do not understand the subject is, that all you official fellows are too fond of needless facts, and are over-tolerant of a redundancy of details.

Milverton. I bow to your greater experience of bores, Ellesmere.

Ellesmere. Again, when you quoted Emerson (not quite accurately, if I remember the passage rightly), I meant to say that it is all very well for him, in his pleasant little Concord, or his all-accomplished Boston,* to make so much of the joys of society; but anything more uncompanionable than the society of London cannot well be imagined.

In these elaborate dinners, and these *réunions*, evening assemblies, or whatever you call them, pleasant companionship is the last thing thought of, or aimed at. In fact, if the metropolis is to go on increasing in size, and society is to be conducted as it is at present, new arrangements will have to be made for friendship and companionship.

---

* This conversation took place some months before the great fire at Boston, or, doubtless, Sir John would have expressed his sorrow at that lamentable catastrophe, and his sympathy with the sufferers.

Sir Arthur. That is an odd phrase, 'arrangements for friendship and companionship.'

Ellesmere. You may disapprove of the phrase if you like; but, depend upon it, the continuance of friendship and the maintenance of companionship can be furthered or hindered by various commonplace things, physical even in their character, which have nothing whatever to do with the romance of friendship or the joys of companionship.

Cranmer. Is not the increase of locomotion a counteracting force, Ellesmere—I mean one that promotes the continuance of friendship?

Ellesmere. Oh, no, not a bit; or, at least, only very little. Just as people see more of the surface of countries by this increased locomotion, so also they may have increased superficial conversancy with former friends and companions. But friendship and companionship must, to a great extent, depend upon neighbourhood. You can't get rid of that physical fact; and therefore the aim of everyone should be to make the necessary daily converse he has with his fellow-creatures a means of friendship and companionship. Look at ourselves. We are all brought together to 'the great wen,' as Cobbett called it—at least for a large part of the year. If I were a country rector, what real use should I be to you for companionship? I think I hear you say, occasionally but very rarely,

'Do you remember that man Ellesmere at college? He is buried somewhere in the country. If he were here now, how he would endeavour to agree with each one of us, and try to place all our views in the best light!' But you would gradually forget me and my merits altogether.

Cranmer. It requires a painful effort of the imagination to picture Ellesmere as a rector in a distant country parish.

Milverton. But there is a great deal of truth in what he says. One must rely much upon neighbourhood, or proximity of some sort, for the maintenance of companionship. Now in London one has no neighbours.

Cranmer. I am going to shock you all; but I must confess that I do like talk of that kind which is vulgarly called 'shop;' and that is a thing independent, to a certain extent, of neighbourhood.

Ellesmere. Very true, Cranmer. 'Shop talk' is very good talk. It is earnest: it is real: and, provided it does not shut other people out of the conversation, it is decidedly a thing to be encouraged rather than discouraged. It is quite right and proper that Cranmer should stand up for it, for of all persons on the earth who talk 'shop,' commend me to official and parliamentary people. A few weeks ago I was honoured by being invited to a male dinner of official and parliamentary persons. I was the only lawyer present, at least the only practising lawyer. I declare to you that, with the exception of myself and my neighbour whom I drew away from his favourite topics, every other person round the table was either talking or listening to official or parliamentary talk. I made the remark at the time, and it was not denied. They all appeared to enjoy themselves amazingly. I therefore am quite prepared to accede to Cranmer's views upon the question.

Now to quite another point. Don't you think it was a little insolent on Milverton's part to speak as he did of the great conversational powers of men of letters?

Milverton. Perhaps it was audacious; but it is true; and the reason of it is not far to seek. It is because their range of subjects is so wide. Now, imagine what a number of questions a man must have looked into who writes a history! I overheard a well-known historian talking the other day on a point of law with one of the most learned of your tribe, Ellesmere. The lawyer said, 'I suppose we are the only two human beings who have looked into this subject for the last century.' I do not unreasonably exalt the conversational merits of men of letters, but only explain them by saying that it is part of their business to know a great many things pretty well. They don't pretend to be finished experts, as professional men are; but they have more general knowledge, which naturally enters into conversation. For instance, the branch of law which that historian was discussing, was perhaps the only one he knew anything about; but probably it gave him an insight into the way you lawyers view questions, and so it put him into a sort of relation towards you all, and made him somewhat of a better companion for you than he would otherwise have been. But, to use a homely proverb, 'the proof of the pudding is in the eating.' What companions many of you must have known among the distinguished men of letters and science of the present day! You ascertain what are the joys of high companionship when you take a walk with Carlyle, Emerson, Froude, Kingsley, Tennyson, Browning, and many others whom I could mention. Then, take those who have gone from us; how great a reputation many of them had as companions, and I do not doubt it was a well-founded reputation! I may name Wordsworth, Sydney Smith, Jeffrey, Rogers, Moore; and, to give an instance of a purely literary man, Dean Milman. That learned dean seemed to his contemporaries to know everything. He was never at fault, they say, in the conversation.

Sir Arthur. Do you remember, in Lockhart's 'Life of Scott,' Sir Walter's coming up on one occasion to London, and being invited day after day to different sets of men, and how he narrates that the conversation of the bishops was the best he heard in London? And why? Because the bishops, as a general rule, are men of letters.

Milverton. I beg to include men of science and men of art in the same category, namely, as having especial qualifications for good companionship. Now, do look at what an artist has to master—nature—man—everything. I have never met with such good judges of character as I have found among artists; and if it were a great object for me to understand the character of any man, I would rather hear what the painter who painted his portrait would say of him, than what even his most intimate friend could tell me.

Sir Arthur. You see, the advantage these men of letters, science, and art, have as companions, is, that their respective occupations make almost every kind of human effort interesting to them. Now, a young engineer was in a company where there was a very learned man. The young fellow talked disrespectfully, as one of the company thought, who was a near relative; (not that he meant to be disrespectful, for any feeling of the kind was foreign from his real nature). This near relative—I may as well say it was his father—took him aside, and said, 'You are very disrespectful, my dear boy; don't you know that he is a man of some celebrity, and I dare say could tell you how to interpret the arrowheaded characters of the time of Sennacherib?' The youth replied, 'Why should I respect him? he can't drive a nail straight through the leathern sole of a highlow; I don't think much of him.' Now where the learned man would 'have a pull,' to use a slang phrase of the day, over this young man is in this way. If it were explained to our learned friend that it was a matter of difficulty to drive a nail straight through a mass of leather, he would probably estimate the feat, and respect the man who could do it accordingly; but it may be well doubted whether the young man would return the compliment, and have an equal estimation for the other's hard work, that of the interpretation of arrow-headed characters.

Ellesmere. After this glorification of men of science, art, and literature, let us hear a little about statesmen. I want to know more about Lord Palmerston. Lord Palmerston's character is a subject upon which Sir Arthur and Milverton, who knew him officially, are always ready to discourse at large. I, too, liked Lord Palmerston hugely, though he was upon the opposite side of the House, and sometimes even dared to make fun of me.

Sir Arthur. At a dinner-table, or in general society, Lord Palmerston, though always genial, did not always shine; but when you were at home with him, or when you were at work with him, or when you were walking with him, he was a charming companion. And what was said about his avoidance of the past, and his intense interest in the present and the future, is exactly true. I saw a great deal of him in the last ten years of his life, and I only remember two or three instances when he went back upon the past; but, as regards our hopes and prospects for the future, he was always ready to discourse at large, whether the subject was agriculture, or law, or politics, or political economy.

Milverton. I am curious to know what were the two or three instances to which you allude.

Sir Arthur. Well : I had the audacity to joke with him about his beautiful handwriting. You know he wrote a better handwriting than any other man of his time. I said, 'I am afraid good handwriting has not always been a sign of great virtue in the writer. The next best handwriting to yours, Lord Palmerston, is Lucrezia Borgia's.' He parried the attack; and then gave instances of good handwriting, and mentioned Louis-Philippe's, describing that monarch's signature to the last document he signed as king—something relating to his abdication, if I recollect rightly. And then Lord Palmerston recounted to me the whole of the secret history of that transaction, and of the king's flight to England.

Milverton. How I should like to have heard it! Give us another instance, Sir Arthur, of his recurrence to the past.

Sir Arthur. I think I have given it before; but, perhaps, not to this company. We were walking together in the grounds at Broadlands, and he began to speak for once, and once only, of his early life.

He told me that early in life his house was next to that of a great lady who frequently gave balls; and he said it used to amuse him to notice the rise, progress, and decline of the ball, as he continued to work throughout the night.

First there were to be heard the faint attempts, at intervals, before many people had arrived, to commence the business of the evening. Then soon afterwards the gaiety began to get warm. Then it became fast and furious. Suddenly there was comparative silence, it being supper-time. Afterwards the gaiety became more fast and more furious. Then it gradually sobered down as the 'small hours' of the night advanced; and the rattling noise of departing carriages began. Then came the last faint accents of the ball, somewhat corresponding with those made by the early attempts at gaiety at the beginning of the evening. Then the last carriage drove away; and the ladies of the next house would come out upon the balcony, looking rather pale and worn. He, too, would go out upon his balcony and watch the dull dawn coming in so mournfully.

Mauleverer. As if it felt **ashamed that** it was about to awaken so many toilworn **and miserable** people to fresh toil and fresh misery.

Sir Arthur. Lord Palmerston did not say anything **like that, Mauleverer.**

While he was talking to me, I kept thinking, not of the dancers, but of himself, and how justly he had attained **to** his great position, by the labour he had given in his earliest years to make himself fitted for it. He was very modest too. Do you remember in the story of his life, how he refused to take a great office and to become a Cabinet Minister when he was yet very young, giving as his reason for refusal, that he was not equal to the place, and had not sufficient knowledge for it?

Ellesmere. I told you that there was no subject upon which we were so sure to make these official men go off, as it were, as the subject of Lord Palmerston.

Milverton. I will tell you how this is. I think he is not sufficiently estimated in the present day. He was a consummate man of business, and that endeared him to me. But, indeed, great statesmen are seldom sufficiently estimated by posterity. How **few** of us know much, comparatively speaking, of Canning, of Huskisson, **of** Lord Aberdeen, and of **many** other statesmen who might be named, and who in fact were very eminent men! They give you the best part of their lives; they are but poorly rewarded during their lifetime; and I think they are not sufficiently remembered afterwards, especially in a free country. The same thing, too, will happen with the present men. The despotic minister of a despotic monarch generally looms large in history; and, whatever his qualifications may have been, he enters into poetry and into history as a prominent figure. Yet the free minister of a constitutional sovereign is often a much greater man, and has much greater difficulties to contend with.

Ellesmere. There is one comfort in talking with Milverton—one always knows where one is with him. The official part of him is never absent.

Milverton. Observe the laboriousness of these men. They deserve ample recognition, if only for that. Oh! I must tell you a story about a man occupying a very different position from a statesman, but you will soon see why this story occurs to me now.

You all know our gardener, John Withers, what a good, kind, cross, tyrannical person he is. Well, there is a certain vegetable of which you have partaken lately, Ellesmere, at our table. From some mere book knowledge of soil and climate, I ventured to say that it would grow in this country, and even in this garden. John **said it would not.** Gardeners, like women, are inveterate **Conservatives.**

Lady Ellesmere. The most useful of men are likely to partake the nature of women.

Milverton. However, feigning an abject humility to his dictum, I still persuaded him, just for the sake of humouring me, to put in some seeds, and attempt to grow this vegetable. It flourished amazingly, as you have seen. My wife (what moral courage, not to say audacity, women possess!) **must** needs comment to **John upon** this success. 'You see, John, it has done very well, as your master said it would.' Then again, after a few minutes, 'Your master was quite right, John.' At length he grew tired of hearing my praises; and he uttered these memorable words:—'*It's not he's saying of it: it's 'tention does it—'tention does it*': meaning that his *at*tention was the real cause of the success. You know his habit of cutting off the first or the last syllable of every important word.

Well, I say too '*It's 'tention does it,*' whether the 'tention is given by gardeners or by statesmen.

The two great requisites for the grandest forms of success are simply to care very much about anything, and to give ''tention' to it.

Sir Arthur. How completely this word ''tention' brings us back to our subject, if indeed it could be said that we have wandered from it! Companions are interesting companions accordingly as they take a deep interest in anything.

Milverton. Almost all persons have a potentiality of good companionship : you have only to get them upon some subject that they really care about, that they have paid ''tention to,' and their companionship, if I may use such a word, is developed.

Sir Arthur. Whenever you enter by some chance into a new phase of society, you find what clever, intelligent, even charming people there are in that society, and what interest they take in matters respecting which, perhaps, you did not think that there was anything interesting at all until you came to know these people. Everywhere there are the elements of good, even of high companionship; and this does not depend upon education. I am afraid the world is beginning to think too much of education. One of the best companions I ever knew was a man who could neither read nor write. He had large views of everything he talked about. Observation and experience had done for him what we are apt to suppose can only be done by a sort of literary education. You will think it perhaps extravagant of me to say; but I seem to myself to see in such men how a Shakespeare arose; for, say what you like, Shakespeare must have had mighty little education, according to our notion of the word.

That makes me think of a weak point in Lord Palmerston. Nothing could persuade him that Shakespeare wrote Shakespeare's Plays. 'What!' he would exclaim, 'you are not so foolish as to suppose that the man you call Shakespeare really did write those plays which show such a wealth of knowledge. I am surprised at you.'

There was then a rather long discussion about Shakespeare's learning, with which my readers need not to be troubled. The conversation, after a time, received quite a different direction, being brought back to its original subject-matter by Sir John Ellesmere's abruptly saying:

A great essay might be written on a subject kindred to your own, Milverton, and which I would entitle 'The unjust claims made for companionship, or the unjust resentments entertained on account of non-companionship or failing companionship.' In the works of all great writers of fiction, I have almost always felt that their greatest successes have been in the portraying of their minor characters. The reason is, that they indulge less in exaggeration in depicting these characters. You will not at first see how this applies. I was thinking that the unjust claims for companionship are generally made by persons who have a spurious kind of humility. Then I naturally thought of Uriah Heep, in one of Dickens's novels; but I soon saw what a much better character, to illustrate my meaning, was that of Mrs. Gummidge, in the same novel. She is always intimating that she is neglected, and her character is admirably drawn. There is a great deal of what I call 'Gummidge' talked in the world, and very unreasonable talk it is.

Sir Arthur. But the whole subject might be much more widely treated. The person who complains of want of companionship nearly always makes an unjust complaint.

Milverton. Yes: he generally fails to throw himself at all into the position of the other side.

For instance, companionship, such as it existed in former days between two persons, is rendered impossible, or at least very difficult, by a change of circumstances; and then one of the companions is nearly sure to blame the other, and to attribute motives, which have nothing whatever to do with the falling off of companionship.

A great mistake, which I suspect we are all prone to make as regards friendship, has been pointed out by a German writer of much originality, named Arthur Schopenhauer; and what he says would apply to companionship. He says, 'What we demand from a friend, and what we promise ourselves on our own behalf, we fix after the measuring-scale of his and our best moments, and thence arises dissatisfaction with others, with ourselves, and with our condition.' *

Now we should never expect from others, nor promise for ourselves, more than belongs to our average capacity in anything.

Ellesmere. I want to carry you away to quite a different subject, in fact to revert to our animal talk, in respect to which there is one point upon which, I think, you have shown yourself a little cowardly, Milverton. I made a note in my mind, to tell you of this.

Milverton. I really don't know what you mean, Ellesmere. I am not in the habit of refraining from saying anything I think upon any subject to any person.

Ellesmere. Well : you have said nothing about sport.

Milverton. If so, it was an oversight. My mind was full of matters connected with the treatment of animals, respecting which I knew something—such as the transit of animals. Moreover, having lived in the

---

* Das, was wir von einem Freunde fordern, und Das, was wir uns von uns selbst versprechen, bestimmen wir nach dem Maaßstab seiner und unserer besten Augenblicke, und daraus erwächst Unzufriedenheit mit Andern, mit uns und mit unserem Zustande.

country, I have been led to observe the management of animals on farms. But, as regards sporting, you could not find a more ignorant human being. My experience of hunting, for instance, which was limited to one occasion, mainly accords with that of Dr. Johnson, who says somewhere :—

I have now learned, by hunting, to perceive that it is no diversion at all, nor ever takes a man out of himself for a moment: the dogs have less sagacity than I could have prevailed on myself to suppose; and the gentlemen often call to me not to ride over them. It is very strange, and very melancholy, that the paucity of human pleasures should persuade us ever to call hunting one of them.

Of course, as you might expect, I agree with most of what Mr. Freeman has written on the subject of sport. And these Pigeon Matches, and things of that kind, seem to me very poor, contemptible, and brutalizing transactions. How women can 'assist' at such entertainments passes my comprehension. But I do not pretend to have thought carefully upon the general subject of sport, and therefore it never occurred to me to dilate upon it. Is that a sufficient answer, Ellesmere?

Ellesmere. H'm! Pretty well: but it is one of those nice questions that I should like to have heard discussed.

Cranmer. And you would have flitted about from

side to side, and would have been equally troublesome to both sides.

Ellesmere. Oh! there is another point in our talk about animals which I wish to put before you. I have been reading this book, Tucker's 'Light of Nature Pursued,' which you told me I was sure to like. You were right. The author is great upon the subject of boys. Listen to what he says about them.

Those of the sprightliest temper are commonly fullest of mischief: because being unable to bear the sight of everything languid around them, they can raise more stir by vexing than by doing service .... and for the like reason they throw stones at people, because they can put them thereby into a greater flutter than by anything else they could do. Or, if they have not an opportunity of seeing the vexation occasioned by the pranks they play, still they can enjoy the thought of it; and will break a window slily, hide a workman's tool, or fasten up his door, for sake of the fancy of how much he will fret and fume when he comes to discover the roguery.*

Then he uses an excellent phrase, namely, 'the engagingness of mischief.' But it was not merely to amuse you that I read these passages. They brought to my mind something which seems to bear upon our general subject. Is not the greatest part of all our cruelty, whether to our fellow-men or to our fellow-

---

* Tucker's *Light of Nature Pursued*, vol. vii. pp. 9, 10.

animals, attributable to our horror of dulness, and our anxiety to 'raise stir'?

Sir Arthur. Your general remark, Ellesmere, is good; but I do not think it has special application to our subject.

Milverton. No; intentional cruelty, from whatever cause arising, is but a small part of the matter. It is careless cruelty, or ignorant cruelty, that we have most to guard against.

Milverton. I will tell you fully my thoughts upon this subject. It appears to me that the great advancement of the world throughout all ages, is to be measured by the increase of humanity and the decrease of cruelty. In hardly anything else do we feel sure that there is advancement. One folly dies out and another folly takes its place. As regards any particular vice or error, it is often difficult to trace any assured improvement for many centuries. For instance, the upper classes drink less imprudently than they did in former generations; but it may be questioned whether, generally speaking, drunkenness is less prevalent or less harmful at present than at any former period. There are times when Art seems to culminate and then to descend. A similar statement would apply to literature. There are golden ages, and then there are silvern; and then there are leaden. Then again there comes, perhaps, a move upwards. Even science, in which there has been a more steady movement of continuous advance than in any other intellectual pursuit, has its periods of comparative decline and elevation. But if you take the history of the world throughout, you must, I think, admit that humanity has been continually upon the increase.

Now you may speak discontentedly about its rate of progress : you may say that desolating wars still are frequent in the world ; but only read the history of former wars, and you will perceive what an immense move has been made in the direction towards humanity, even in these most barbarous transactions. I am convinced that if an historian of world-wide knowledge, supposing there could be such a man, were to sum up the gains and losses of the world at the end of each recorded century, there might be much which was retrograde in other aspects of human life and conduct, but nothing that could show a backward course in humanity. Even as regards those instances of cruelty and brutality which have been mentioned in the course of our conversations, it must be admitted that some are exceptional ; that others are heedless rather than intentional ; and that when brought before the bar of public opinion, they are universally condemned. It would also, I think, be seen in any such review of a century as I have imagined, that all which was supported by inhumanity, or even by violence, has come to nought; and as that **great poet** Schiller says **of a tumultuous** inundation, the destruction **alone which it** has caused remains to **show that it ever existed.*** A river is not formed: **nothing that is useful for men remains** as the **result of all this violence, and we can** only trace its existence **by the** mischief **it has accomplished.**

It seems **as if the main** design of Providence had **been to bring upon this earth a race** of beings **perpetually improving as regards humanity.** We arrogate this word **to ourselves.** I cannot but believe, that if we ceased **to** fulfil **the conditions which are assumed in that word, we should be supplanted by another race.** And therefore I feel, **that even in this minor** matter (if indeed it **can be called a minor matter)** of our treatment **of the inferior animals** associated with us, if we failed **to exercise the requisite humanity, it would** go hard with ourselves. **The fact that our own** material welfare depends **much upon our treatment of**

---

* Reißen die Brücken und reißen die Dämme
Donnernd mit fort im Wogengeschwemme,
Nichts ist, das die Gewaltigen hemme;
Doch nur der Augenblick hat sie geboren:
Ihres Laufes furchtbare Spur
Geht verrinnend im Sande verloren,
Die Zerstörung verkündigt sie nur.
        Die Braut von Messina.—Schiller.

the lower animals, rather seems to verify than to militate against the foregoing conclusion.

Ellesmere. I have not been so much interested for a long time in any subject as I am in this. When it was first started, I expected from the nature of our host, that it would be treated rather sentimentally, and though in this case I happen to sympathise with the sentiment, I did not expect to hear much of what was practically interesting; but I have come to consider this matter as one of the deepest importance at the present time. During your absence, Milverton has told me many things which he has not mentioned in general conversation. You see, with the exception of certain official and scientific persons, he ought to know more than most people about the treatment of cattle.

Milverton. It is not very pleasant to be talked about in this way before one's face; but still, as the result is not unpleasing, I accept the talk. It means that Ellesmere thinks I am in general given to theory and sentiment; but that on this rare occasion I have something practical to say for myself.

Ellesmere. I flatter myself that I also have found something very practical to bring before you; but I must do so by degrees, working up to my point. Milverton has imparted to me several of the main facts relating to cattle plague. He showed me from data that cannot be contradicted, that in the great outbreak of '66 we lost cattle, the value of which amounted to more than five millions. He then showed me that the number of animals lost during the present year in Europe amounted to several hundreds of thousands. You see this is a very serious thing for the whole world.

Milverton. I wish here to recall to your memories a fact to which you may have given little heed at the time. During the late war between France and Germany, you may occasionally have noticed that a foreign correspondent would say, 'The French peasants in this neighbourhood declare that they have suffered more from the cattle plague introduced by the Germans, than by any depredations committed by the German soldiery.'

Ellesmere. Milverton also showed me how vast had been the destruction of cattle, from the disease, during two periods in the course of the last century.

Cranmer. Well, we admit all this—what then?

Ellesmere. Do you remember that at the beginning of our talk — the first day we were all here — Sir Arthur was rather put down by Milverton for bringing forward difference of race as a cause of any difference of treatment of animals? This branch of the subject ought, in my opinion, to have had more weight given to it. Can anyone tell us anything of this difference of treatment? What about Russia, for instance?

Sir Arthur. All that I know of the treatment of animals in Russia is derived from Custine. I should not say that his idea of the treatment is favourable. And he makes this observation, that the Russians have no laws for the protection of animals, as there are in England. What made me notice this passage was that I was surprised to find that Custine had any knowledge of our laws.

Milverton. It is such a wide subject, that one is afraid of being dogmatic about it, and of being dogmatically wrong. I don't mind giving you my impressions; but I have scarcely anything more than impressions to offer. From what I have heard and seen the Germans are very good to their animals. So are the French, with the exception of those abominable vivisecting surgeons. In Spain, if you recollect, Ellesmere, we observed great brutality in the treatment of the mules, in the diligences; but otherwise we thought the Spaniards tolerably humane to animals. I make a special exception, of course, in regard to their bull-fights. You see, this is a subject which probably has seemed to travellers not worthy of much observation. Nothing that I have read in books of travel in America has given me much information as regards their treatment of animals. I am very curious to know what it is; and if I were to travel in America, it would be one of the principal things that I should try to ascertain. I also long to know how animals are treated in Russia. These two nations are, together with ourselves, the most growing nations in the world; and the future treatment of animals must greatly depend upon the views and habits affecting these creatures, which may be adopted by the Americans, the Russians, and ourselves. I can venture to contradict Custine on this point. There is a law which certainly prevails in some of the Governments of Russia, that when you are posting, you must have a horse for every adult passenger. If there are six of you, you must hire six horses, not necessarily attached to the same carriage, but three horses to each set of three travellers, for they would divide your party into two. This is a small matter, but still it shows some effort at legislation in favour of animals.

Ellesmere. Now you are coming close to the point which is, in my mind, of the greatest practical importance and which I have been working up to.

You have admitted that this cattle plague is one of the most serious things at present before us. I can even foresee that serious political troubles might be occasioned by the dearness of provisions, which this pest may at some time suddenly produce.

Now, here is the question. Is it the treatment of

animals, or is it climate, or some other unavoidable cause, which breeds this disease? Suppose it to be treatment; suppose it to be something which more judicious, or rather less injudicious mode of housing, feeding, and wintering could mitigate or prevent—how all important it becomes to Europe, and indeed, to the whole world, to ascertain these facts! I am, of course, alluding to those parts of the earth where this disease is supposed to be bred.

Here is something practical to be done. Galicia, Milverton tells me, is a country from which, as a centre, cattle plague spreads. He says, that he knows of but one man who has travelled in Galicia, and that good man, Professor Symonds, went with the object of investigating the cattle plague. His knowledge has been most serviceable to this country. But I want a great deal more done. I want some one to go there and to stay and study the agricultural modes of proceeding in that country, and in those parts of Russia, where the cattle plague may be almost said to have its home. Your friend Ruskin, Milverton, is often showing men what they should do to make life better and more beautiful. I wish we could enlist his eloquence on our side. Here really is a great field for wise exertion.

Cranmer. I declare Ellesmere is becoming quite enthusiastic and eloquent.

Ellesmere. If you don't behave yourself, Cranmer, I shall suggest that the Government should do something, and that I know will give you pain, for it will cost money.

Now, do you all follow me? I do not mind repeating concisely what I wish to urge upon you. This cattle plague is a tremendous evil. Its continual presence is to be dreaded by reason, as Milverton told me, of the increase of railway facilities for transit. The railway system is now complete from those districts which are the supposed foci of the disease to our own shores. We should try to master the evil at these foci. You who care very much about the sufferings of animals, should hate this disease, as it is one accompanied by great suffering to the poor creatures. But even more than that you should care, if you are statesmen, for the supply of food to the poorest of our population. It may be that the treatment of the animals at these central places of infection is the cause of the infection; and if so, the remedy consists in improving that treatment there. If cholera were the subject of our discourse, I should put forth similar views about that. Here is an object of real benevolence of the most practical kind. Milverton said, we might be satisfied that our talk had not been useless, if we put down the bearing-rein; but what is the bearing-rein compared with the cattle plague? The object that I have put forward, embraces almost every branch of the subject we have been considering—the instruction to be given about animals; the treatment of animals by the owner; the transit of animals; the effects of difference of race in the treatment of animals; and indeed, every part of the subject that has direct practical importance, both as regards humanity, policy, and political economy. I have said my say.

Sir Arthur. I wish, Ellesmere, you had said all this, before I gave my summary of our sayings. I assure you, I should not have neglected to give your suggestion a prominent place.

Milverton. As you may imagine, I agree with every word that Ellesmere has said. I only wish to point out to you a further consequence that may result from the investigations he has proposed. It is this, that if it should be proved that the treatment of animals in those countries which are the foci of this disease, should not be the true cause of its existence being developed there, a system of quarantine combined with slaughter must be adopted jointly by all the principal countries of Eastern Europe.

Mrs. Milverton. Now I should like to go back to companionship. Nothing has been said of the companionship of men and women. I mean nothing special relating to that branch of the subject.

Lady Ellesmere. Very true, Blanche. I observed that when these men went into a discussion of the heights of grand companionship, it was only the society of other men they were thinking of. Rather unpolite, I thought.

Milverton. By way of answer to you both, the words of two poets immediately occur to me. Please give me 'Philip Van Artevelde,' Johnson.

In that beautiful interlude between the two parts of Philip's life, there is the following passage. I must first, though, give you a little description of the context. Elena, early in life, before she met with Philip, had fallen in love with a young man very inferior to herself :—

> The well-spring of his love was poor
> Compared to hers : his gifts were fewer ;
> The total light that was in him
> Before a spark of hers grew dim ;
> Too high, too grave, too large, too deep,
> Her love could neither laugh nor sleep—
> And thus it tired him.

Then the poet goes on to say—

> His desire
> Was for a less consuming fire :
> He wished that she should love him well,
> Not wildly ; wished her passion's spell
> To charm her heart, but leave her fancy free ;
> To quicken converse, not to quell ;
> He granted her to sigh, for so could he ;
> But when she wept, why should it be ?
> 'Twas irksome, for it stole away
> The joy of his love holiday.

Now I have always thought that Sir Henry Taylor has treated that young man, name unknown, rather harshly, for I strongly suspect that most men, even of the superior kind, greatly resemble him. I do not think that Elena could have been a very pleasant companion, and I believe that companionship has a great deal to do with the maintenance of love. The love is to be a thing always understood, coming under the first requisite mentioned in the essay as the basis of companionship, namely, personal liking; but it does not require to be perpetually expressed. Now I come to the other poet, Byron. Manfred says, I recollect the words :—

> She was like me in lineaments—her eyes,
> Her hair, her features, all, to the very tone
> Even of her voice, they said were like to mine;
> But soften'd all, and temper'd into beauty;
> She had the same lone thoughts and wanderings,
> The quest of hidden knowledge, and a mind
> To comprehend the universe; nor these
> Alone, but with them gentler powers than mine,
> Pity, and smiles, and tears—which I had not,
> And tenderness—but that I had for her;
> Humility—and that I never had.
> Her faults were mine—her virtues were her own.

These two passages give two instances of the companionship of men and women—one successful, the other the reverse. The exceeding charm of this companionship will always consist in difference of character. In the latter instance, the difference co-exists with a tender similitude. That, perhaps, would create the highest form of companionship.

In the former instance, what was wanting, the want of which troubled this small-souled young man as it would trouble most men, was the kind of love which should ensure good companionship, fitted

> To quicken converse, not to quell.

The difference of character should have given a zest to companionship, not destroyed it. To create this kind of companionship great intellectual gifts are not required on the part of the lesser personage of the two companions, but only perceptiveness and receptiveness; though, of course, it will always be the better and the larger if there should be these intellectual gifts. I have nothing more to say upon the subject.

Lady Ellesmere. You have not said anything, Leonard, about men and women who live together, and who differ upon every earthly subject. How will they get on?

Milverton. I really think, my dear, I have met this case. I said that difference rather created pleasure than not, in companionship, and that this might even extend to difference of tastes; but that the difference must not be made offensive. You all know what I mean. One cannot state such complicated matters in such a way as to guard them from all objection. We all know, especially you ladies, when we are making ourselves disagreeable, and so embittering companionship. There are times when a man will bear to hear his most cherished convictions not only opposed, but ridiculed. There are other times when it is unsafe to do this. Tact, of which you women possess an almost unfair share, will easily enable you to distinguish between these times; and your large possession of this invaluable quality, tact, enables you to become at all times the most delightful companions that can be imagined to us poor men.

Lady Ellesmere. Let us go immediately, Blanche. I have a wild desire, have not you? to see how your good, cross-grained gardener, John, who has had so much weight and influence in this conversation, has planned out the new gravel-walk; which we must all say is the closest approach to perfection. If we stay any longer here, my husband or some of the other gentlemen will begin taking exceptions, or making qualifications which will tend to deface and disfigure the fine things which Leonard has just been saying of us. Not that I quite like what he said about Elena; but still I suppose it was as much as we could expect from a man.

The ladies immediately rose, we followed them; and the rest of the afternoon was spent in considering gardening projects, submitted with the greatest deference, even by Sir John, to his namesake, the gardener.

What an advantage a bad temper would be to one if it did not bite at both ends; if it did not make one's self nearly as miserable as it does other people!

This was the last day of our holiday, and we returned on the morrow to our weary work in London.

INDEX.

ABERDEEN, Lord, 186
Accuracy, a divine, 35
Addison, quotation from, 143
Adverse argument, stating an, 69
Agreements, family, 28
Agriculture, Bastiat on, 11
Anaxagoras on animal reason, 64, 84
Angelic Doctor, quotation from the, 88
Animals, courtesy to, 101
— cruelty to, 8, 19, 43, 44, 45, 193, 195, &c.
— diseases of, 11
— dumb, 129
— familiarity with, 9
— fear, 48
— memory of, 57
— overcrowding, 18
— poets and, 120
— reasoning powers of, 65, 108
— rights of, 44
— sermons on behalf of, 20
— sufferings of, 99

Animals (*continued*), Sunday rest for, 63
— treatment of, in Russia, 200; from habits of race, 12, 199
Anecdote about calumny, 25
— Mr. Cranmer and his father, 31
— Ellesmere and the drowning cat, 131
— young engineer and man of letters, 183
— Frenchman out riding, 22
— fox and hounds, 55
— Milverton's upset from boat, 3
— the two statesmen in a cab, 161
— Thackeray and oysters, 50
— Lord Palmerston, 183-6, 190
— John Withers, 187
— the word 'contrive,' 145
Anxiety, 49
Apology for Raimond de Sebonde. Quotation from Montaigne's, 91
Applause, 142
Aquinas, Thomas, quotation from, 88
'Arcana Politica,' quotation from the, note, 86
Aristotle, 75; on man, 105
Arnold, Dr., quotation from, 98
Arnold, Edwin, quotation from, 108
Arrowheaded characters, 183
Art of writing, the, 33
Artists as judges of character, 182
Avitus, Junius, 29

BACON, quotation from, 140; and Machiavelli, 138
Balls, Lord Palmerston's account of a neighbour's, 185
Barclay, Dr., quoted by Grindon, 100

Barlæus's anecdote of a fox, 55
Barrow's Sermons, quoted for clearness of expression, 34
Bastiat, 10; quotation from, 11
Bayle, quotation from, 98
Bearing-reins, 10, 66; not used in Scotland, 67; human, 72, 78
Beavers, 52
Berkeley, Bishop, on designing, 84; on law-making, 85
Birds, as pets, 23; Voltaire on, 74; Montaigne on, 91; Petrarch on, 93
Bishops as companions, 185
Blair, Dr., 70
Bores, 174–76
Boyle on animals, 102
Boys, and accidents, 4; Tucker on, quotation from 'Light of Nature Pursued,' 194
Breaks, want of, 80
Brougham, quotation from Lord, 118
Browning as a companion, 181
Burke as a good companion, 175
Butterflies, ladies and, 50
'Bus horse's letter in the 'Echo,' 80
Byron, quotation from, 206

CAB, anecdote of two statesmen in a, 161
Caged birds, 23
Calumny, anecdote about, 25
Canning but little known, 186
Cardan, Jerome, quotation from, 86
Care and anxiety, 49
Carlyle as a companion, 181
Cathedrals, origin of, 139

Catholics admitted to colleges, 85
Cato, M., quotation from Plutarch's life of, 101
Cattle, transit of, 15
— plague, the, 199
Cats and their masters, 25
— love for music, 88
— drowning a, 131
Central control, office for, 77
Cervantes, 124
Character, artists good judges of, 182
Cholera, 203.
Christianity and modern preaching, 20
Churchill, quotation from, 144
Civilisation and horses, 21
Clearness of expression, 34
Cobbett on London, 178
Colleges, admitting Catholics to, 85
Collie-dogs, 24
'Common-place Book of Thoughts,' quotation from Mrs. Jameson's, 99
Communication with cab-masters, 61
Companionship, essay on, 163. Universality of, 164. Fascinating people, 164. Shakespeare on, 164. Emerson, 165. Personal liking and friendship, 166. Von Humboldt, 167. Steadfastness, 168. Talking and silence, 168. Difference of temperament, 170. Unreasonable expectations of some people, 171. Talking of the past, 173. Lord Palmerston, 173, 184. Bores, 174-176. Statesmen, 174. Burke, 174. London, 177-179. Bishops, 182. Education, 189. Schopenhauer on, 191
'Compitum,' quotation from Digby's, 124
Consultation among friends, 5
'Contrive,' 142, 145

Control, central, 77
Conversations, good done by, 6
Cortes and his treatment of the Indians, 57
Courtesy to animals, 101
Coutts, the Baroness Burdett, 159
Cranmer, Mr., and his father, 31
Criticism, hostile, 146
Cruelties to animals, 8. Inflicted by hirelings, 19. In ancient Rome, 92. Inspections as a preventative, 45. Woman's influence over, 43. Public opinion, 44. Spaniards, 125. Pigeon matches, 193. Careless and ignorant, 195
Culture and familiarity, 9 ; and the bearing-rein, 13-14
Custine on treatment of animals in Russia, 201

DE QUINCEY, 148
Dean, Richard, quoted by Grindon, 100
Descartes, 63, 73, 88, 125
Designing, Bishop Berkeley on, 84 ; on law-making, 85
Details and boredom, 176
Dialogue, its uses and drawbacks, 7
'Dialogues on Instinct,' quotation from, 118
Dickens, 27, 191
'Dictionnaire philosophique,' quotation from, 75
Digby's 'Compitum,' quotation from, 124
Dinners, wearisomeness of long, 71. Dogs and, 23
'Discours de la Méthode,' 73
Diseases of Animals, 11
Divine accuracy, 35
Dobson's translation of Petrarch, 93
Dogs, 23. 'Fairy,' 24. Collies, 24. Fuller on, 83
Double entry and Politics, 147
Dumb animals, 129

INDEX.

'ECHO,' 'bus horse's letter in, 80
Education, 189; and familiarity, 9; and the bearing-rein, 31–14
Ellesmere's anecdote of the cat and Milverton, 131
'Eloge' of Voltaire on Descartes, 73
Emerson, 165; and companionship, 177, 181
Employers, 61
Engineer, anecdote of the man of letters and the young, 183
Epictetus, 142
Exceptions, admitting, 90

'FAIRY,' Mr. Milverton's dog, 24, 54, 65
Fascinating people, 164
Familiarity with animals, 9
Family agreements, 28
Fathers and sons, 29
Faust, quotation from, 49
Fear, sufferings of animals from, 48, 52
Fines, imposition of, on conquered nations, 137
Fish, gold-, official life and, 24
Fox, Barlæus's anecdote of a, 55
Freeman, Mr., 193
Frenchman, anecdote of the, riding, 22
Friends in council, list of the, 1
Friendship and companionship, 165
Froude, Mr., 181
Fun, dogs' enjoyment of, 56
Fuller, 93

GALEN, 104
Gardeners and conservatives, 188

Geese, a case of cruelty to, 113
Geheimrath, a faulty, 19
German papers, obscurity in, 36
Gibbon, 144
Glass bowl, fish in a, 24
Goethe, 34, 49, 64, p.
Gold-fish, 24
Governmental action, 17
'Guardian,' quotation from, 121
Gulliver, 22
'Gummidgery,' 191

HALF-WEEKLY post, 38
'Harmonies Economiques,' 11
Heartbeating, 105
Herbert, George, quotation from the life of, 123
Hirelings, cruelty inflicted by, 19
High-companionship, essay on the joys of, 163
Horses and warfare, 20; and civilisation, 21; noiseless when hurt, 21; timidity of, 59; shying, 59; Montaigne on, 91
Hume, David, quotation from, 108
Hudibras, quotation from, 101
Huskisson, 186

IMAGINATION and cruelty, 57
Immortality of the soul, 96
— of brutes, 100
Improving away a likeness, 7
Inspections to prevent cruelty to animals, 45
Intellectual victory, love of, 70
Interference, State, 77, 79
Ironclads and cathedrals, 139

JAMESON, Mrs., quotation from, 98, 99
　Jeffery, 181
Johnson, Dr., quotation from, 193 ; saying about a boat, 3
Juvenile presumption, 30

KINGSLEY, 181
　Koran, 157

LAME horses, 44
　Language, clearness of, 34, 76
Law-making, 85
Lavater and expression, 69
Legislators, born, 85
Letter from ''Bus horse' in the 'Echo' newspaper, 80
— writing, 38
Lewes, G., and penguins, 104
Liberty, personal, 77
Licensing Act, working of the new, 78
'Light of Nature Pursued,' quotation from, 119
Likenesses improved away, 7
Locke, 121
London society, 177
Loss to a nation through want of proper care for animals, 11
Love of parents, 30
Love-making, Sir John Ellesmere's ideas upon, 6
Lucid language, 34, 76
Lucretius, quotation from, 107
Lytton, Lord, 27

MACHIAVELLI, quotation from, 140; and Bacon, 138
　'Macmillan's Magazine,' 78
Matter and soul, 89

Medecin, and cruelty towards animals, 41
Meddygon Myddfai, 41
Memory of cruelty, 57
Men of letters good companions, 181
Men's minds, travelling over, 72
Middle passage, cruelties of the, 18
Mill, John, 77
Milman, Dean, 181
Milverton, Mr., upset from a boat, 3; rescues a cat from drowning, 131
Montaigne, quotations from, p., 91, 92. Essay on cruelty, 91. Apology for Raimond de Sebonde, 91. Referred to by Pope, 121
Moore, 181
Morin, 95
Mothers' and fathers' feelings, 29
Mulcts, national, 137
Munro's translation of Lucretius, 107

NEIGHBOURS, 179
'Night Thoughts,' quotation from, 96
'Nobody does nothing,' 26
Noise, making a, when hurt, 21
Nonsense, suiting people's, 150
North, quotation from Sir Thos., Plutarch's 'Lives,' 100

OBJECTIONS and exceptions, 90
Official routine, 24
Omnibus breaks, 80
Over-crowding, 18, 112
Ovid, 104, 121
Oysters, Thackeray and the, 50

PALEY'S definition of virtue, 98
Palmerston, Lord, 173, 183-6; and Shakespeare, 190
Pamphleteering, a dead thing, 6
Partnerships in writing, 37
Past, talking of the, 173
Penguins, 102, 104
Pereira, Gomez, 124
Petrarch, quotation from, 93, 121
Pets, 23, 24
Philip v. Artevelde, quotation from, 205
Pigeon matches, 193
Pleasure-taking, 63
Pliny, quotation from, 29
Plutarch, 84, 100
Poets and animals, 120
Political talkers, 180
Post, half-weekly, 38
Posting in Russia, 201
Pope, quotation from, 121
Popularity, Machiavelli on, 140
Power, loss of time in deciding where to be placed, 45
— of reason in animals, 65, 108
Prices increased by overcrowding stock, 18
Providence and nature, 65
Public documents, writing, 36
— opinion, effect of, on cruelties, 44
Pythagoras, 91

'QUERIST,' Bishop Berkeley's, 84
Quotations from authors :—
Addison, 143
Aquinas, Thos., 88

Quotations from authors—*continued*
    Aristotle, 105
    Arnold, Dr., 98
    Arnold, Edwin, 108
    Bacon, 140
    Bastiat, 11
    Bayle, 98
    Bentham, J., 99
    Berkeley, 84
    Brougham, 118
    Byron, 206
    Cardan, J., 86
    Churchill, 144
    Emerson, 165
    Fuller, 93
    Galen, 104
    Grindon, 100
    Goethe, p. 65
    Hume, 108
    Jameson, Mrs., 98, 99
    Johnson, Dr., 193
    Lavater, 69
    Machiavelli, 140
    Montaigne, 91, 92
    Morin, 95
    Munro, 107
    Nash, 107
    North, 101
    Ovid, 104
    Petrarch, 93
    Pliny, 29
    Plutarch, 101
    Pope, 121

Quotations from authors—*continued*
    Racine, 104
    Ramage, 167
    Richter, 153
    Ruskin, p.
    Schiller, 137, 197,
    Schopenhauer, 191
    Seneca, 87
    Spinoza, 106
    Steele, 122
    Tennyson, 164
    Tucker, 119, 194
    Uhland, 125
    Voltaire, 74
    Walton, 123
    Young, 96
    Xenophanes, 108
From books, &c. :—
    'Beautiful Thoughts from German Authors,' 167
    Bible, 67, 157
    'Bride of Messina,' 197
    'Common Book of Thoughts,' 99
    'Dialogues on Instinct,' 118
    'Dictionnaire philosophique,' 75
    'Dream of Fair Women,' 164
    'Echo,' 80
    'Faust,' 49
    'Guardian,' 121
    'Harmonies Economiques,' 11
    'Hesperus,' 153
    'Hudibras,' 102
    Koran, 157
    'Life of George Herbert,' 123

Quotations from books, &c.—*continued*
    'Life, its Nature, Varieties, and Phenomena,' 100
    'Light of Nature Pursued,' 194
    'Lucretius,' 107
    Petrarch's 'Lives,' 101
    'Poets of Greece,' 108
    'Querist,' 84
    Racine, 104
    'Revelation of St. Bridget,' 124
    St. François d'Assise, 95
    'Summa Totius Theologiæ,' 88
    'Tatler,' 122
    'Tractatus Teologico-politicus,' 106
    'Van Artevelde, P.,' 205
    'View of Human Life,' 93
Quarantine, International, 204

RACE, influence in the treatment of animals from, 12, 199
Racine, quotation from, 104
Ramage's 'Beautiful Thoughts from German Authors,' 167
Reasoning powers of animals, 65, 108
Rebelliousness of young people, 29
Regulations and liberty, 77
'Remain, I do not,' 22
'Reminds me that,' 52
Revelations of St. Bridget, 124
Richter, J. P., 153
Ridicule, evil effects of, 115
Rights of animals, 44–46
Rochefoucauld, want of a, among horses, 21
Romans and public cruelties, 92
Rogers, 181

Ruskin, p. and 202
Russia, treatment of animals in, 200

ST. BRIDGET, Revelations of, 124
St. Francis d'Assise, 94
St. Hilaire, Isidore Geoffroy, on penguins, 105
Scamps, 30
Scandal, 25
Schiller, 36 ; quotation from, 137
Schopenhauer, 191
Scotch carters and the bearing-rein, 68
Scott, Sir W. in London, 180
Seneca, quotation from, 87
Sermons about animals, 20
Service, good, to be requited, 94
Shakespeare—regard for animals, 95 ; as a companion, 164 ; Lord Palmerston and, 190
Sheep, loss of, from overcrowding, 15
Ships, cattle-carrying, 15
Shying horses, 59–60
Silence over grievances, 21
Sir Arthur restates arguments, 127
Smith, Sydney, 181 ; pamphlets, 6
Sons and fathers, 28
Sorrel nag, the, 22
South Kensington Museum anticipated, 84
Space to be allotted to beasts during transit, 116
Spaniards and cruelty, 125
Sport, 192
Subject of this book, choice of, 1
Sufferings of animals, 99
'Summa Totius Theologiæ,' 88

Summary of conversation, Sir Arthur's, 127
Sunday rest for beasts, 63
Superstition, 41
Stafford's, Lord, letter to Walpole, 92
State interference, 77
— papers, writing, 36
Steadfastness, 168
Steele, quotation from, 122
Storms, effect of, on animals, 95
Symonds, Prof., 202

TACT, 208
     Talking of the past, 173; and silence, 168
'Tatler,' quotation from, 122
Taylor, Sir H., 206
Temper, bad, 209
Tenderness to old age, 93
Tennyson, quotation from, 164; as a companion, 181
''Tention,' 188
Thackeray's wit, 50
Theatrical performances, influences from, 122
Themes, writing, 37
Theories of Descartes, 88; respecting soul and matter, 89
Theorists, 86
Timidity of the horse, 59
Transit Committee, 14, 159
— of animals, 14, 116
'Tractatus Theologico-politicus,' 103, 105
Travelling over people's minds, 72

UHLAND, quotation from, 125

VARTEVELDE, quotation from, 205
Virtue, Paley's definition of, 98
'View of Human Life,' quotation from, 93
Vivisection, 43. Voltaire on, 75
Voltaire and Descartes, 73
Von Humboldt and companionship, 167

WALPOLE, Horace, letter to Lord Stafford, 92; and Montaigne, 92
Walton, Isaac, 123
War, 20, 136
Water, life compared to boiling, 86
Waterton's love for animals, 54
Withers, J., 187, 208
Women's influence in preventing cruelty, 43, 117, 208
Wordsworth, 181
Writing clearly, 33; Lord Palmerston's, 186
— and criticism, 146

XENOPHANES, quotation from, 108

YOUNG people, Pliny on, 29
Young's 'Night Thoughts,' 96

LONDON: PRINTED BY
SPOTTISWOODE AND CO., NEW-STREET SQUARE
AND PARLIAMENT STREET

BOOK LIST

ABLE TO SAVE; or Encouragement to Patient Waiting. By the Author of "The Pathway of Promise." Cloth antique, 2s. 6d.

ABOUT'S (EDMOND) Handbook of Social Economy; or the Worker's A B C. Crown 8vo, 5s.

ACKWORTH VOCABULARY, or English Spelling Book; with the Meaning attached to each Word. Compiled for the use of Ackworth School. New Edition. 18mo, 1s. 6d.

ADAMS' (W. H. DAVENPORT) Famous Ships of the British Navy; or Stories of the Enterprise and Daring of British Seamen. With Illustrations. Crown 8vo, cloth gilt extra, 3s. 6d.

ÆSOP'S FABLES. With 100 Illustrations, by Wolf, Zwecker, and Dalziel. Square 32mo, cloth gilt extra, 2s. 6d.; neat cloth, 1s. 6d.

AIDS TO PRAYER. Cloth Antique, 1s. 6d.

ALEXANDER'S (BISHOP) The Divine Death. A Sermon preached in St. Paul's Cathedral on Good Friday, 1871. Sewed, 1s.

ALFORD'S (DEAN) The Book of Genesis and Part of the Book of Exodus. A Revised Version, with Marginal References and an Explanatory Commentary. Demy 8vo, 12s.

———— The New Testament. Authorised Version Revised. Long Primer, crown 8vo, 6s.; Brevier, fcap. 8vo, 3s. 6d.; Nonpareil, small 8vo, 1s. 6d., or in calf extra 4s. 6d.

Strahan & Co.'s

ALFORD'S (DEAN) Essays and Addresses, chiefly on Church Subjects. Demy 8vo, 7s. 6d.

——————— The Year of Prayer; being Family Prayers for the Christian Year. Crown 8vo, 3s. 6d.; small 8vo, 1s. 6d.

——————— The Week of Prayer. An Abridgment of "The Year of Prayer;" intended for use in Schools. Neat cloth, 9d.

——————— The Year of Praise; being Hymns, with Tunes, for the Sundays and Holidays of the Year. Large type, with music, 3s. 6d.; without music, 1s. Small type, with music, 1s. 6d.; without music, 6d. Tonic Sol-fa Edition, crown 8vo, 1s. 6d.

——————— How to Study the New Testament. Part I. The Gospels and the Acts.—II. The Epistles (first section).—III. The Epistles (second section) and the Revelation. Small 8vo, 3s. 6d. each.

——————— Eastertide Sermons. Small 8vo, 3s. 6d.

——————— The Queen's English. A Manual of Idiom and Usage. New and Enlarged Edition. Small 8vo, 5s.

——————— Meditations: Advent, Creation, and Providence. Small 8vo, 3s. 6d.

——————— Letters from Abroad. Crown 8vo, 7s. 6d.

——————— Poetical Works. New and Enlarged Edition. Crown 8vo, 5s.

——————— Biblical Revision: Its Duties and Conditions. A Sermon preached in St. Paul's. Sewed, 1s.

——————— The Compacted Body. A Sermon preached at the Consecration of the Suffragan Bishop of Dover. Sewed, 1s.

ANDERSEN'S (HANS CHRISTIAN) The Will-o'-the-Wisps are in Town; and other New Tales. With Illustrations. Square 32mo, 1s. 6d.

ANDREWS' (REV. S. J.) The Bible-Student's Life of Our Lord. Crown 8vo, 3s. 6d.

ARGYLL'S (THE DUKE OF) The Reign of Law. Crown 8vo, 6s. People's Edition, limp cloth, 2s. 6d.

————— Primeval Man. An Examination of some Recent Speculations. Crown 8vo, 4s. 6d.

————— Iona. With Illustrations. Crown 8vo, Crown 8vo, 3s. 6d.

BARTLETT'S (W. H.) Walks about the City and Environs of Jerusalem. With 25 Steel Engravings and numerous Woodcut Illustrations. 4to, cloth gilt extra, 10s. 6d.

BASKET OF FLOWERS; or Piety and Truth Triumphant. A Tale for the Young. By G. T. Bedell, D.D. 32mo, 1s.

BAUERMAN'S (H., F.G.S.) A Treatise on the Metallurgy of Iron. Post 8vo, 12s.

BAUR'S (WILLIAM) Religious Life in Germany during the Wars of Independence, in a series of Historical and Biographical Sketches. New and cheaper Edition. Crown 8vo, 7s. 6d.

BAYNE'S (PETER) Life and Letters of Hugh Miller. Two Vols., demy 8vo, 32s.

————— The Days of Jezebel. An Historical Drama. Crown 8vo, 6s.

BEACH'S (CHARLES) Now or Never; or the Trials and Perilous Adventures of Frederick Lonsdale. Crown 8vo, cloth gilt extra, 3s. 6d.

BEECHER'S (HENRY WARD, D.D.) Prayers in the Congregation. Crown 8vo, 3s. 6d.

————— Eyes and Ears. Crown 8vo, 3s. 6d.

————— Life Thoughts. Small 8vo, 2s. 6d.

————— Royal Truths. Crown 8vo, 3s. 6d.

————— Pleasant Talk about Fruits, Flowers, and Farming. Small 8vo, 2s. 6d.

BELLENGER'S One Hundred Choice Fables, in French, imitated from La Fontaine. With Dictionary of Words and Idiomatic Phrases, Grammatically Explained. New Edition, revised by C. J. Delille. 12mo, 1s. 6d.

BENJIE OF MILLDEN. By the Author of "Bygone Days in our Village." Sewed, 6d.

BENNOCH'S (FRANCIS, F.S.A.) Sir Ralph de Rayne and Lilian Grey. Small 8vo, 1s. 6d.

BENONI BLAKE, M.D. By the Author of "Peasant Life in the North." Two Vols., crown 8vo, 21s.

BEVERLEY'S (MAY) Romantic Tales from English History. New Edition, with 21 Illustrations. Crown 8vo, cloth gilt extra, 3s. 6d.

BJÖRNSON'S (BJORNSTJERNE) Arne; a Sketch of Norwegian Peasant Life. Translated by Augusta Plesner and Susan Rugeley-Powers. Crown 8vo, 5s.

BLACKIE'S (J. S.) Lays of the Highlands and Islands. Small 8vo, 6s.

BLAIKIE'S (W. G., D.D.) Better Days for Working People. Crown 8vo, boards, 1s. 6d.

——— Counsel and Cheer for the Battle of Life. Crown 8vo, boards, 1s. 6d.

——— Heads and Hands in the World of Labour. Crown 8vo, 3s. 6d.

——— The Head of the House. Sewed, 2d.

BOARDMAN'S (REV. W. E.) The Higher Christian Life. Small 8vo, 9d.

BOOTH'S (E. CARTON) Another England. Victoria. Post 8vo, 3s. 6d.

——— Homes away from Home and the Men who make them in Victoria. Demy 8vo, 6d.

BOWLES' (CAROLINE) Harmless Johnny. Sewed, 3d.

BRADY'S (W. MAZIERE, D.D.) Essays on the English State Church in Ireland. Demy 8vo, 12s.

BRAMSTON'S (MARY) Cecy's Recollections. A Story of Obscure Lives. Crown 8vo, cloth gilt extra, 5s.

BRITISH SPORTS AND PASTIMES. Edited by Anthony Trollope. Post 8vo, 10s. 6d.

BROWN'S (JOHN, M.D.) Plain Words on Health. Lay Sermons to Working People. Sewed, 6d.

BROWN'S (J. E. A.) Lights Through a Lattice. Small 8vo, 3s. 6d.

———————— Palm Leaves. From the German of Karl Gerok. Cloth antique, 6s.

BROWNE'S (MATTHEW) Views and Opinions. Crown 8vo, 6s.

BUCHANAN'S (ROBERT) Idyls and Legends of Inverburn. Crown 8vo, 6s.

———————————— London Poems. Crown 8vo, 6s.

———————————— Undertones. Small 8vo, 6s.

———————————— The Book of Orm. Crown 8vo, 6s.

———————————— Napoleon Fallen. A Lyrical Drama. Crown 8vo, 3s. 6d.

———————————— The Drama of Kings. Post 8vo, 12s.

———————————— The Fleshly School of Poetry. Crown 8vo, sewed, 2s. 6d.

BÜCHSEL'S (REV. DR.) My Ministerial Experiences. Crown 8vo, 3s. 6d.

BULLOCK'S (REV. CHARLES) The Way Home; or the Gospel in the Parable. Small 8vo, 1s. 6d.

BUSHNELL'S (HORACE, D.D.) Moral Uses of Dark Things. Crown 8vo, 6s.

———————————— Christ and His Salvation, in Sermons variously related thereto. New and Cheaper Edition. Crown 8vo, 4s. 6d.

———————————— Christian Nurture; or the Godly Upbringing of Children. Crown 8vo, 3s. 6d.

———————————— Nature and the Supernatural, as Together constituting the One System of God. Crown 8vo, 3s. 6d.

———————————— The Character of Jesus. Limp cloth, 6d.

———————————— The New Life. Crown 8vo, 3s. 6d.

BUSHNELL'S (Horace, D.D.) The Vicarious Sacrifice, grounded on Principles of Universal Obligation. Crown 8vo, 7s. 6d.

―――――― Work and Play. Crown 8vo, 3s. 6d.

CAIRNS' (John, D.D.) Romanism and Rationalism, as opposed to Pure Christianity. Sewed, 1s.

CAMDEN'S (Charles) When I was Young. A Book for Boys. With Illustrations. Crown 8vo, cloth gilt extra, 2s. 6d.

―――――― The Boys of Axleford. With Illustrations. Crown 8vo, cloth gilt extra, 5s.

CAPES' (Rev. J. M.) Reasons for Returning to the Church of England. Crown 8vo, 5s.

CARLILE'S (Rev. J., D.D.) Manual of the Anatomy and Physiology of the Human Mind. Crown 8vo, 4s.

CARTWRIGHT (Peter, the Backwoods Preacher), Autobiography of. Edited by W. P. Strickland. Crown 8vo, 2s.

CHILD WORLD. By the Authors of "Poems written for a Child." With Illustrations. Square 32mo, cloth gilt extra, 3s. 6d.

CHILD NATURE. By one of the Authors of "Child World." With Illustrations. Square 32mo, cloth gilt extra, 3s. 6d.

CHILD WORLD LIBRARY. Containing—

| Child World. | Child Nature. |
| Stories Told to a Child. | Poems Written for a Child. |

Four square 32mo vols., profusely illustrated, and handsomely bound, in neat case, 16s.

CHILDREN'S JOURNEY (The), &c. By the Author of "Voyage en Zigzag." With Illustrations. Square 8vo, 10s. 6d.

CHRISTIAN COMPANIONSHIP FOR RETIRED HOURS. Crown 8vo, cloth gilt extra, 3s. 6d.

CHURCH LIFE: Its Grounds and Obligations. By the Author of "Ecclesia Dei." Crown 8vo, 2s. 6d.

COLONIES (THE) AND IMPERIAL UNITY. By the Author of "Ginx's Baby." Sewed, 1s.

CONTEMPORARY REVIEW (THE): Theological, Literary, and Social. 2s. 6d. monthly. Four-Monthly Volumes, 10s. 6d. each.

COOLIE (THE): His Rights and Wrongs. Notes of a Journey to British Guiana, with a Review of the System, and the Recent Commission of Inquiry. By the Author of "Ginx's Baby." Post 8vo, 16s.

COX'S (REV. SAMUEL) The Resurrection. Crown 8vo, 5s.

—— The Private Letters of St. Paul and St. John. Crown 8vo, 3s.

—— The Quest of the Chief Good. Expository Lectures on the Book Ecclesiastes, with a New Translation. Small 4to, 7s. 6d.

CRAIG'S (ISA) Duchess Agnes, and other Poems. Small 8vo, 5s.

CRITICAL ENGLISH TESTAMENT (THE); Being an Adaptation of Bengel's Gnomon, with numerous Notes, showing the Precise Results of Modern Criticism and Exegesis. Edited by Rev. W. L. Blackley, M.A., and Rev. James Hawes, M.A. Complete in Three Volumes, averaging 750 pages. Crown 8vo, 6s. each.

CUPPLES' (MRS. GEORGE) Tappy's Chicks, and other Links between Nature and Human Nature. With Illustrations. Crown 8vo, cloth gilt extra, 5s.

DAILY DEVOTIONS FOR CHILDREN. 32mo, 1s. 6d.

DAILY MEDITATIONS FOR CHILDREN. 32mo, 1s. 6d.

DALE'S (R. W.) Week-Day Sermons. Crown 8vo, 3s. 6d.

DALTON'S (WM.) Adventures in the Wilds of Abyssinia; or The Tiger Prince. With Illustrations. Crown 8vo, cloth gilt extra, 3s. 6d.

DAVIES' (EMILY) The Higher Education of Women. Small 8vo, 3s. 6d.

DE BURY'S (Baroness Blaze) All for Greed. Popular Edition. Crown 8vo. [*Nearly Ready.*

DE GASPARIN'S (Countess) Human Sadness. Small 8vo, 5s.

———————— The Near and the Heavenly Horizons. Crown 8vo, 3s. 6d.

———————— A Poor Boy. Sewed, 6d.

DE GUERIN'S (Eugenie) Journal. Crown 8vo, cloth gilt extra, 5s.

———————— Letters. Crown 8vo, 5s.

DE LIEFDE'S (John) The Charities of Europe. With Illustrations. Crown 8vo, 5s.

———————— The Postman's Bag. A Story Book for Boys and Girls. With Illustrations. Crown 8vo, cloth gilt extra, 3s. 6d.

———————— Days of Grace. With Illustrations. Crown 8vo, 5s.

———————— The Pastor of Gegenburg, and other Stories. With Illustrations. Crown 8vo, 5s.

DE WITT'S (Madame, née Guizot) A French Country Family. Translated by the Author of "John Halifax." With Illustrations. Crown 8vo, cloth gilt extra, 5s.

DENISON'S (E. B., LL.B., Q.C., F.R.A.S., &c.) Life of Bishop Lonsdale. Crown 8vo, 2s. 6d.

DICKSEE'S (J. R.) School Perspective. A Progressive Course of Instruction in Linear Perspective. Post 8vo, 5s.

DISCUSSIONS ON COLONIAL QUESTIONS; being the Report of the Proceedings of a Conference held at Westminster Palace Hotel. Crown 8vo, 2s. 6d.

DOBNEY'S (Rev. H. H.) Free Churches. Post 8vo. 4s. 6d.

———————— A Vision of Redemption. Sewed, 4d.

DODD'S (G.) Dictionary of Manufactures. Post 8vo, 5s.

DRESSER'S (C.) Unity in Variety, as deduced from the Vegetable Kingdom. With Illustrations. 8vo, 10s. 6d.

———————— Rudiments of Botany, Structural and Physiological; being an Introduction to the Study of the Vegetable Kingdom. With numerous Illustrations. 8vo, 15s.

DU LYS' (COUNT VETTER) Irma. A Tale of Hungarian Life. Two Vols., post 8vo, 18s.

DUPANLOUP'S (MGR., Bp. of Orleans) Studious Women. Translated by R. M. Phillimore. Crown 8vo, 4s.

DUTCHMAN'S (A) Difficulties with the English Language. Sewed, 6d.

ECCLESIA DEI: The Place and Function of the Church in the Divine Order of the Universe, and its Relations with the World. Demy 8vo, 7s. 6d.

EIGHT MONTHS ON DUTY. The Diary of a Young Officer in Chanzy's Army. With a Preface by C. J. Vaughan, D.D., Master of the Temple. Crown 8vo, 5s.

ENGLAND'S DAY. A War Saga. Commended to Gortschakoff, Grant, and Bismarck, and Dedicated to the British Navy. Sewed, 6d.

EPISODES IN AN OBSCURE LIFE. Crown 8vo, 6s.

EVENINGS AT THE TEA TABLE. With Illustrations. Uniform with "Stories told to a Child." Square 32mo, cloth gilt extra, 3s. 6d.

FAIRHOLT'S (F. W.) Dictionary of Terms in Art. With numerous Illustrations. Post 8vo, 6s.

FERNYHURST COURT. An Every-day Story. By the Author of "Stone Edge." Crown 8vo, 6s.

FIELD'S (GEORGE). The Rudiments of Colours and Colouring. Revised, and in part rewritten, by Robert Mallet, M.A., F.A.S., &c. With Illustrations. Crown 8vo, 4s. 6d.

FITZGERALD'S (PERCY) Proverbs and Comediettas, written for Private Representation. Crown 8vo, 6s.

FLETCHER'S (Rev. Alexander, D.D.) Assembly's Catechism. Divided into Fifty-two Lessons. 12mo, sewed, 8d.

FOUNDATIONS OF OUR FAITH (The). By Professors Auberlen, Gess, and others. Crown 8vo, 3s. 6d.

FRANKLIN'S (John) Illustrations to the Ballad of St. George and the Dragon. Small 4to, cloth gilt extra, 10s. 6d.

FRASER'S (Rev. R. W., M.A.) The Seaside Naturalist: Out-door Studies in Marine Zoology and Botany, and Maritime Geology. With Illustrations. Crown 8vo, cloth extra, 3s. 6d.

FRIENDS AND ACQUAINTANCES. By the Author of "Episodes in an Obscure Life." Crown 8vo, 6s.

FRIENDLY HANDS AND KINDLY WORDS. Stories illustrative of the Law of Kindness, the Power of Perseverance, and the Advantages of Little Helps. Crown 8vo, cloth gilt extra, 3s. 6d.

GARRETT'S (Edward) Occupations of a Retired Life. Crown 8vo, 6s.

———— The Crust and the Cake. Crown 8vo, 6s.

———— Premiums Paid to Experience. Incidents in my Business Life. Two Vols., post 8vo, 21s.

———— Seen and Heard. Three Vols. Post 8vo.

GEIKIE'S (Cunningham, D.D.) Life. A Book for Young Men. Crown 8vo, cloth extra, 3s. 6d.

———— Light from Beyond, to Cheer the Christian Pilgrim. Cloth antique, 2s. 6d.

GERHARDT'S (Paul) Spiritual Songs. Translated by John Kelly. Small square 8vo, 5s.

GILBERT'S (William) De Profundis. A Tale of the Social Deposits, Crown 8vo, 6s.

———— Doctor Austin's Guests. Crown 8vo, 6s.

GILBERT'S (WILLIAM) The Magic Mirror. A Round of Tales for Old and Young. With Illustrations. Square 32mo, cloth gilt extra, 2s. 6d.

———— King George's Middy. With Illustrations. Crown 8vo, cloth gilt extra, 6s.

———— The Washerwoman's Foundling. With Illustrations. Square 32mo, cloth gilt extra, 2s. 6d.

———— The Wizard of the Mountain. Two Vols., post 8vo, 21s.

———— Shirley Hall Asylum. New Edition. Crown 8vo, 10s. 6d.

GILES' English Parsing. Improved Edition. 12mo, 2s.

GLADSTONE'S (THE RIGHT HON. W. E.) On "Ecce Homo." Crown 8vo, 5s.

GOOD WORDS. Edited by Donald Macleod, 6d. monthly, Illustrated. Yearly Volumes, 1860 to 1872. Cloth gilt extra, 7s. 6d. each.

———— First Set of Ten Volumes, 1860 to 1869. Uniformly bound in cloth, extra gilt and black, £3 15s.

The Volumes are not sold separately in this Binding.

GOOD WORDS FOR THE YOUNG. Edited by George MacDonald, LL.D. 6d. monthly, Illustrated. Yearly Volumes, 1869 to 1871. Cloth gilt extra, 7s. 6d. each.

GOSSE'S (PHILIP HENRY, F.R.S.) A Year at the Shore. With Thirty-six Illustrations, printed in Colours. Crown 8vo, 9s.

GOTTHELF'S (JEREMIAH) Wealth and Welfare. Crown 8vo, 6s.

GREENWELL'S (DORA) Essays. Crown 8vo, 6s.

———— Poems. Enlarged Edition. Crown 8vo, 6s.

———— The Covenant of Life and Peace. Small 8vo, 3s. 6d.

GREENWELL'S (DORA) The Patience of Hope. Small 8vo, 2s. 6d.

——————— Two Friends. Small 8vo, 3s. 6d.

——————— Colloquia Crucis. Small 8vo, 3s. 6d.

——————— On the Education of the Imbecile. Sewed, 1s.

GREGORY'S (BENJAMIN) The Thorough Business Man. Memoirs of Walter Powell, Merchant, of Melbourne and London. With Portrait. Crown 8vo, 6s.

GUTHRIE'S (THOMAS, D.D.) Early Piety. 18mo, 1s. 6d.

——————— Man and the Gospel. Crown 8vo, 3s. 6d.

——————— Our Father's Business. Crown 8vo, 3s. 6d.

——————— Out of Harness. Crown 8vo, 3s. 6d.

——————— Speaking to the Heart. Crown 8vo, 3s. 6d.

——————— Studies of Character from the Old Testament. First and Second Series. Crown 8vo, 3s. 6d. each.

——————— The Angels' Song. 18mo, 1s. 6d.

——————— The Parables Read in the Light of the Present Day. Crown 8vo, 3s. 6d.

——————— Sundays Abroad. Crown 8vo, 3s. 6d.

HACK'S (MARIA) Winter Evenings; or Tales of Travellers. With Illustrations. Small 8vo, cloth gilt extra, 3s. 6d.

——————— Grecian Stories. With Illustrations. Small 8vo, cloth gilt extra, 3s. 6d.; smaller Edition, 2s. 6d.

HALL'S (MR. and MRS. S. C.) Book of the Thames, from its Rise to its Fall. With Fourteen Photographic Illustrations and One Hundred and Forty Wood Engravings. Fcap. 4to, cloth gilt extra, 21s.

HARE'S (AUGUSTUS J. C.) Walks in Rome. Two Vols., crown 8vo, 21s.

——— Memorials of a Quiet Life. Two Vols., crown 8vo, 21s.

——— Wanderings in Spain. With Illustrations. Crown 8vo, 10s. 6d.

HARGREAVES' (JOHN GEORGE) The Blunders of Vice and Folly, and their Self-acting Chastisements. Crown 8vo, 6s.

HARRIS (SIR W. SNOW, F.R.S.) A Treatise on Frictional Electricity, in Theory and Practice. Edited, with Memoir, by Charles Tomlinson, F.R.S. 8vo, 14s.

HAWEIS' (REV. H. R.) Music and Morals. Post 8vo, 12s.

HAWTHORNE'S (NATHANIEL) Passages from English Note-books. Edited by Mrs. Hawthorne. Two Vols., post 8vo, 24s.

——————— Passages from French and Italian Note-books. Two Vols., post 8vo, 24s.

HENRY HOLBEACH: Student in Life and Philosophy. A Narrative and a Discussion. With Letters to Mr. M. Arnold, Mr. Alexander Bain, Mr. T. Carlyle, Mr. A. Helps, Mr. G. H. Lewes, Rev. H. L. Mansel, Rev. F. D. Maurice, Mr. J. S. Mill, and Rev. Dr. J. H. Newman. Second Edition, with Additions. Two Vols., post 8vo, 14s.

HEROINES OF THE HOUSEHOLD. By the Author of "The Heavenward Path," &c. With Illustrations. Crown 8vo, cloth gilt extra, 3s. 6d.

HERSCHEL'S (SIR J. F. W., BART.) Familiar Lectures on Scientific Subjects. Crown 8vo, 6s.

HOGE'S (REV. W. J.) Blind Bartimeus and his Great Physician. Small 8vo, 1s.

HOLBEACH'S (HENRY) Shoemakers' Village. Two Vols., crown 8vo, 16s.

HOLMES' (OLIVER WENDELL) The Autocrat of the Breakfast Table. With Illustrations. Small 8vo, 3s. 6d.

HOOPER'S (Mrs.) Recollections of Mrs. Anderson's School. A Book for Girls. With Illustrations. Small 8vo, 3s. 6d.

HOWE'S (Edward) The Boy in the Bush. With Illustrations. Crown 8vo, cloth gilt extra, 5s.

HOWSON'S (Dean) The Metaphors of St. Paul. Crown 8vo, 3s. 6d.

———— Proportion in Religious Belief and Religious Practice. A Sermon preached at the Consecration of the Bishop of Carlisle. Sewed, 1s.

———— The Companions of St. Paul. Crown 8vo, 5s.

———— The Character of St. Paul. Crown 8vo. [*In the Press.*

HUNT'S (Rev. John) History of Religious Thought in England, from the Reformation to the End of Last Century. Vols. I. and II., demy 8vo, 21s. each.

———— An Essay on Pantheism. Post 8vo, 10s. 6d.

HUNTINGTON'S (F. D., D.D.) Christian Believing and Living. Crown 8vo, 3s. 6d.

HUTTON'S (R. H.) Essays, Theological and Literary. Two Vols., square 8vo, 24s.

HYMNS FOR THE YOUNG. With Music by John Hullah. 8vo, 1s. 6d.

INGELOW'S (Jean) Mopsa the Fairy. With Illustrations. Crown 8vo, cloth gilt extra, 5s.

———— Studies for Stories. With Illustrations by Millais and others. Crown 8vo, cloth gilt extra, 5s.

———— A Sister's Bye-hours. With Illustrations. Cloth gilt extra, 5s.

———— Stories Told to a Child. With Illustrations. Square 32mo, cloth gilt extra, 3s. 6d. Also in eight separate books. Neat cloth, 6d. each.

IRVING'S (Edward) Collected Writings. Five Vols., demy 8vo, £3.

IRVING'S (EDWARD) Miscellanies from the Collected Writings. Post 8vo, 6s.

——————— Prophetical Writings. Vols. I. and II., demy 8vo, 15s. each.

JACOB'S (G. A., D.D.) The Ecclesiastical Polity of the New Testament. A Study for the Present Crisis in the Church of England. Post 8vo, 16s.

JOHNSTONE'S (REV. J. BARBOUR) "It is Your Life." Preaching for the People. Crown 8vo, 2s. 6d.

JONES (AGNES ELIZABETH) Memorials of. By her Sister. With a Portrait. Crown 8vo, 3s. 6d.

JONES' (ARCHDEACON) The Peace of God. Crown 8vo, 5s.

JONES' (REV. HARRY, M.A.) The Regular Swiss Round. With Illustrations. Small 8vo, 3s. 6d.

KAYE'S (JOHN WILLIAM) Lives of Indian Officers, illustrative of the History of the Civil and Military Service of India. Three Vols., crown 8vo, 6s. each.

KERR'S (JOHN) Lessons from a Shoemaker's Stool. Sewed, 6d.

KINGSLEY'S (REV. CHARLES) Madam How and Lady Why. With Illustrations. Square 8vo, cloth gilt extra, 7s. 6d.

——————— Town Geology. Crown 8vo, 5s.

KINGSLEY'S (HENRY) The Boy in Grey. With Illustrations. Crown 8vo, 3s. 6d.

KINGSTON'S (W. H. G.) Foxholme Hall, and other Amusing Tales for Boys. With Illustrations. Small 8vo, cloth gilt extra, 3s. 6d.

——————— The Pirate's Treasure, and other Amusing Tales for Boys. With Illustrations. Small 8vo, cloth gilt extra, 3s. 6d.

——————— Harry Skipwith. A Tale for Boys. With Illustrations. Small 8vo, cloth gilt extra, 3s. 6d.

KRILOF AND HIS FABLES. By W. R. S. Ralston. With Illustrations. New and much enlarged Edition. Crown 8vo, 7s. 6d.

LEE'S (Rev. F. G., D.C.L.) The Christian Doctrine of Prayer for the Departed. With copious Notes and Appendices. Demy 8vo, 16s.

LEGENDS OF KING ARTHUR AND HIS KNIGHTS OF THE ROUND TABLE (The). Compiled and Edited by J. T. K. Small 8vo, sewed, 1s.; cloth, 1s. 6d.

LEIGHTON'S (Robert) The Laddie's Lamentation on the Loss of his Whittle, and other Poems. Sewed, 6d.

LEITCH'S (William, D.D.) God's Glory in the Heavens. With Illustrations. Crown 8vo, 4s. 6d.

LE PAGE'S FRENCH COURSE.
"The sale of many thousands, and the almost universal adoption of these clever little books by M. Le Page, sufficiently prove the public approbation of his plan of teaching French, which is in accordance with the natural operation of a child learning his native language."

French School. Part I. L'Echo de Paris. A Selection of Familiar Phrases which a person would hear daily if living in France. 12mo, 3s. 6d.

N.B. A Key to the above, being Finishing Exercises in French Conversation. 18mo, 1s.

—————————. Part II. The Gift of Fluency in French Conversation. 12mo, 2s. 6d.

N.B. A Key to the above: "Petit Causeur; or, First Chatterings in French." 12mo, 1s. 6d.

—————————. Part III. The Last Step to French. With the Versification. 12mo, 2s. 6d.

Petit Lecteur des Colléges; or, the French Reader, for Beginners and Elder Classes. A Sequel to "L'Echo de Paris." 12mo, 3s. 6d.

French Master for Beginners; or, Easy Lessons in French. 12mo, 2s. 6d.

Juvenile Treasury of French Conversation. With the English before the French. 12mo, 3s.

Ready Guide to French Composition. French Grammar by Examples, giving Models as Leading-strings throughout Accidence and Syntax. 12mo, **3s.** 6d.

Etrennes aux Dames Anglaises. A Key to French Pronunciation in all its niceties. Sewed, 6d.

LILLIPUT LEVEE. Poems of Childhood, Child-fancy, and Child-like Moods. With Illustrations by Millais and others. Square 32mo, cloth gilt extra, 2s. 6d.

LILLIPUT LECTURES. By the Author of "Lilliput Levee." With Illustrations. Square 8vo, cloth gilt extra, 5s.

LILLIPUT LEGENDS. By the Author of "Lilliput Levee." With Illustrations. Square 8vo, cloth gilt extra, 5s.

LLOYD'S (Mrs. W. R.) The Flower of Christian Chivalry. With Thirty-four Illustrations by J. D. Watson and others. Crown 8vo, cloth gilt extra, 3s. 6d.

LOCKER'S (Frederick) London Lyrics. Small 8vo, 6s.

LOSSING'S (Benson J.) The Hudson from the Wilderness to the Sea. Illustrated by 300 Engravings on Wood. Small 4to, cloth gilt extra, 21s.

LOVING COUNSEL. An Address to his Parishioners. By the Author of "The Pathway of Promise." Limp cloth, 8d.

LUDLOW'S (J. M.) Woman's Work in the Church. Small 8vo, 5s.

LUDLOW (J. M.) and LLOYD JONES' The Progress of the Working Class from 1832 to 1867. Crown 8vo, 2s. 6d.

LYNCH'S (Rev. T. T.) Sermons for my Curates. Edited by the Rev. Samuel Cox. Post 8vo, 9s.

———— Letters to the Scattered. Post 8vo, 9s.

———— The Rivulet. A Contribution to Sacred Song. New Edition. Small 8vo, 3s. 6d.

———— Tunes to Hymns in "The Rivulet." Edited by T. Pettit, A.R.A.M. Square 8vo, 2s. 6d.

LYTTELTON'S (Hon. and Rev. W. H.) Ordination Sermon preached in Worcester Cathedral, 1871. Sewed, 1s.

MAC DONALD'S (GEORGE) Annals of a Quiet Neighbourhood. Crown 8vo, 6s.

——————— The Seaboard Parish. Crown 8vo, 6s.

——————— Wilfrid Cumbermede. Crown 8vo, 6s.

——————— Dealings with the Fairies. With Illustrations by Arthur Hughes. Square 32mo, cloth gilt extra, 2s. 6d.

——————— The Disciple and other Poems. Crown 8vo, 6s.

——————— Unspoken Sermons. Crown 8vo, 3s. 6d.

——————— The Miracles of our Lord. Crown 8vo, 5s.

——————— The Wow o' Rivven. Sewed, 6d.

——————— At the Back of the North Wind. With Illustrations. Crown 8vo, cloth gilt extra, 7s. 6d.

——————— Ranald Bannerman's Boyhood. With Illustrations. Crown 8vo, cloth gilt extra, 5s.

——————— The Princess and the Goblin. With Illustrations. Crown 8vo, cloth gilt extra, 5s.

——————— Works of Fancy and Imagination: being a reprint of Poetical and other Works. Pocket-volume Edition, in neat case, £2 2s.

MAC DONALD'S (MRS. GEORGE) Chamber Dramas for Children. Crown 8vo, 7s. 6d.

MACKAY'S (CHARLES) Studies from the Antique, Sketches from Nature, and other Poems. Small 8vo, 3s. 6d.

MACKENZIE and IRBY'S (MISSES) Travels in the Sclavonic Provinces of Turkey in Europe. With Illustrations. Demy 8vo, 24s.

Book List.

MACLEOD'S (NORMAN, D.D.) Peeps at the Far East. With Illustrations. Small 4to, cloth gilt extra, 21s.

——————— Eastward. With Illustrations. Crown 8vo, 6s.

——————— Character Sketches. With Illustrations. Post 8vo, 10s. 6d.

——————— Thoughts on the Temptation of Our Lord. Crown 8vo.

——————— Job Jacobs and his Boxes. In packets of Thirteen, 1s.

——————— Parish Papers. Crown 8vo, 3s. 6d.

——————— Reminiscences of a Highland Parish. Crown 8vo, 6s.

——————— Simple Truth spoken to Working People. Small 8vo, 2s. 6d.

——————— The Earnest Student: being Memorials of John Mackintosh. Crown 8vo, 3s. 6d.

——————— The Gold Thread. A Story for the Young. With Illustrations. Square 8vo, 2s. 6d.

——————— The Old Lieutenant and his Son. With Illustrations. Crown 8vo, 3s. 6d.

——————— The Starling. With Illustrations. Crown 8vo, 6s.

——————— Wee Davie. Sewed, 6d.

——————— How can we best Relieve our Deserving Poor? Sewed, 6d.

——————— Concluding Address to the Assembly of the Church of Scotland. May, 1869. Sewed, 1s.

——————— War and Judgment. A Sermon preached before and published by command of the Queen. Sewed, 1s.

MANSEL'S (Dean) The Philosophy of the Conditioned: Sir William Hamilton and John Stuart Mill. Post 8vo, 6s.

MANUAL OF HERALDRY: being a Concise Description of the several Terms used, and containing a Dictionary of every Designation in the Science. Illustrated by 400 Engravings on Wood. Small 8vo, 3s.

MARKBY'S (Rev. Thomas) Practical Essays on Education. Crown 8vo, 6s.

MARLITT'S (E.) The Old Maid's Secret. Translated by H. J. G. Crown 8vo, 6s.

MARSH'S (J. B.) The Story of Harecourt; being the History of an Independent Church. With an Introduction by Alexander Raleigh, D.D. With Illustrations. Crown 8vo, 6s.

MARSHMAN'S (J. C.) Story of the Lives of Carey, Marshman, and Ward. Crown 8vo, 3s. 6d.

MARTIN'S (Rev. H.) The Prophet Jonah. Crown 8vo, 6s.

MARTIN'S (W.) Noble Boys. Their Deeds of Love and Duty. With Illustrations. Crown 8vo, cloth gilt extra, 3s. 6d.

MASSEY'S (Gerald) A Tale of Eternity, and other Poems. Crown 8vo, 7s.

MAURICE'S (Rev. F. D.) The Working Man and the Franchise; being Chapters from English History on the Representation and Education of the People. Demy 8vo, 7s. 6d.; crown 8vo, boards, 1s. 6d.

MAZZINI'S (Joseph) The War and the Commune. Sewed, 1s.

MERIVALE'S (Charles, B.D., D.C.L.) Homer's Iliad. In English Rhymed Verse. Two Vols., demy 8vo, 24s.

METEYARD'S (Eliza) The Doctor's Little Daughter. The Story of a Child's Life amidst the Woods and Hills. With Illustrations. Crown 8vo, cloth gilt extra, 5s.

MILLAIS' ILLUSTRATIONS. A Collection of Drawings on Wood. By John Everett Millais, R.A. Demy 4to, cloth gilt extra, 16s.

MONRO'S (Rev. Edward) Edwin's Fairing. With Illustrations. Square 32mo, cloth gilt extra, 2s. 6d.

NEILL'S (Edward D.) The English Colonization of America during the Seventeenth Century. Demy 8vo, 14s.

NEWMAN'S (John Henry, D.D.) Miscellanies from the Oxford Sermons, and other Writings. Crown 8vo, 6s.

NOEL (The Hon. Roden) The Red Flag and other Poems. Small 8vo, 6s.

NUGENT'S (E., C.E.) Optics; or Sight and Light Theoretically and Practically Treated. With numerous Woodcuts. New and Enlarged Edition. Post 8vo, 5s.

NUTTALL'S (Dr.) Dictionary of Scientific Terms. Post 8vo, 5s.

ORACLES FROM THE BRITISH POETS. A Pleasant Companion for a Round Party. By James Smith. Fourth Edition. Small 8vo, 2s. 6d.; or in antique, morocco gilt, 5s.

ORME'S (Benjamin) Treasure Book of Devotional Reading. Crown 8vo, cloth gilt extra, 3s. 6d.

OSBORN'S (Rev. H. S., M.A.) The Holy Land, Past and Present. Sketches of Travel in Palestine. With Fifty Illustrations on Wood and Steel. Crown 8vo, cloth gilt extra, 3s. 6d.

OUR COMMON FAITH. Popular Expositions of the Apostles' Creed. By Eminent Ministers of various Sections of the Church. Small 8vo. [*In preparation.*

PARKES-BELLOC'S (Bessie Rayner) Essays on Woman's Work. Small 8vo, 4s.

PARKES-BELLOC'S (BESSIE RAYNER) La Belle France. With Illustrations. Square 8vo, 12s.

——————————— Vignettes. **Twelve** Biographical Sketches. Crown 8vo, 6s.

PARKINSON'S (J. C.) A Day at Earlswood. Sewed, 6d.

PARR'S (MRS.) Dorothy **Fox.** Crown 8vo, 6s.

—————— How it all Happened, and other Stories. Two Vols., post 8vo, 21s.

PARRY (CHARLES, Commander Royal Navy) Memorials of. By his Brother, the Right Rev. Edward Parry, D.D., Suffragan Bishop of Dover. Crown 8vo, 5s.

PATHWAY OF PROMISE (THE). Cloth antique, 1s. 6d.

PATTIE DURANT. A Tale of 1662. By the Author of "Passing Clouds," &c. Small 8vo, 2s. 6d.

PAUL GOSSLETT'S CONFESSIONS IN LOVE, LAW, AND THE CIVIL SERVICE. With Illustrations by Marcus Stone. Post 8vo, 2s. 6d.

PEASANT LIFE IN THE NORTH. New and Cheaper Edition. Crown 8vo, 6s.

————————————————————— Second Series. Crown 8vo, 9s.

PEEPS AT FOREIGN COUNTRIES. With Illustrations. Crown 8vo, cloth, gilt extra, 5s.

PEROWNE'S (REV. CANON) The Athanasian Creed. Sewed, 1s.

PERSONAL PIETY. A Help to Christians to Walk worthy of their Calling. Cloth antique, 1s. 6d.

PHELPS' (AUSTIN) Man's Renewal. Small 8vo, 2s. 6d.

—————— The Still Hour. Small 8vo, 1s.

PHILLIMORE'S (John George) History of England during the Reign of George the Third. Vol. I., 8vo, 18s.

PICTORIAL SPELLING-BOOK; or Lessons on Facts and Objects. With 130 Illustrations. New Edition. 12mo, 1s. 6d.

PLUMPTRE'S (Professor) Biblical Studies. Post 8vo, 7s. 6d.

——————— Christ and Christendom; being the Boyle Lectures for 1866. Demy 8vo, 12s.

——————— Lazarus and other Poems. Crown 8vo, 5s.

——————— Master and Scholar, and other Poems. Crown 8vo, 5s.

——————— Sunday. Sewed, 6d.

——————— The Tragedies of Æschylos. A New Translation, with a Biographical Essay, and an Appendix of Rhymed Choruses. Two Vols., crown 8vo, 12s.

——————— The Tragedies of Sophocles. A New Translation, with a Biographical Essay, and an Appendix of Rhymed Choruses. Crown 8vo, 7s. 6d.

——————— Theology and Life. Sermons chiefly on Special Occasions. Small 8vo, 6s.

——————— "The Spirits in Prison." A Sermon on the state of the Dead. Preached in St. Paul's Cathedral. Sewed, 1s.

POEMS WRITTEN FOR A CHILD. By Two Friends. With Illustrations. Square 32mo, cloth gilt extra, 3s. 6d.

PORTER'S (Noah, D.D.) The Human Intellect, with an Introduction upon Psychology and the Soul. Demy 8vo, 16s.

——————— The Elements of Intellectual Science. A Manual for Schools and Colleges. Demy 8vo, 10s. 6d.

PRESENT-DAY PAPERS on Prominent Questions in Theology. Edited by the Right Rev. Alexander Ewing, D.C.L., Bishop of Argyll and the Isles. One Shilling each; or in Three Vols., crown 8vo, 7s. 6d. each.

    I. THE ATONEMENT.
    II. THE EUCHARIST.
    III. THE RULE OF FAITH.
    IV. THE PRESENT UNBELIEF.
    V. WORDS FOR THINGS.
    VI. PRAYERS AND MEDITATIONS.
    VII. JUSTIFICATION BY FAITH.
    VIII. MOTHER-CHURCH.
    IX. USE OF THE WORD REVELATION IN THE NEW TESTAMENT.
    X. THE CHRISTIAN MINISTRY. Part 1.
    XI. THE CHRISTIAN MINISTRY. Part 2.
    XII. THE ETERNAL LIFE MANIFESTED.
    XIII. SOME LETTERS OF THOMAS ERSKINE OF LINLATHEN.
    XIV. GOD AND THE CHRISTIAN SACRAMENTS.
    XV. ST. AUGUSTINE AND HIS MOTHER.
    XVI. SOME FURTHER LETTERS OF THOMAS ERSKINE OF LINLATHEN.
    XVII. THE FUTURE TEMPORAL SUPPORT OF THE MINISTRY.
    XVIII. THE RELATION OF KNOWLEDGE TO SALVATION.
    XIX. RECONCILIATION.

PREVAILING PRAYER. With Introduction by Norman Macleod, D.D. Crown 8vo, 1s. 6d.

PRITCHARD'S (REV. CHARLES) The Testimony of Science to the Continuity of the Divine Thought for Man. A Sermon preached at the Meeting of the British Association for 1869. Sewed, 1s.

RALEIGH'S (ALEXANDER, D.D.) The Little Sanctuary. Crown 8vo, 6s.

—————— When our Children are about us. Sewed, 3d.

REED (ANDREW, D.D.), Memoirs of the Life and Philanthropic Labours of. By his Sons. With Portrait and Illustrations. Crown 8vo, 6s.

RIPPON'S (Dr.) Selections of Hymns from the Best Authors, including a great number of Originals, intended as an Appendix to Dr. Watts' Psalms and Hymns.

Nonpareil 32mo.—Roan, 1s. 6d.; Roan, gilt edges, 2s.
Long Primer, 24mo.—Roan, 2s. 6d.; Roan, gilt edges, 3s.
Large Type.—Sheep, 5s.; Roan, gilt edges, 6s.

ROBERTSON'S (John, D.D.) Sermons and Expositions. Post 8vo, 7s. 6d.

ROBINSON CRUSOE. With Illustrations. 18mo, cloth gilt extra, 1s. 6d.

ROGERS' (Henry) Essays from "Good Words." Small 8vo, 5s.

SABINE'S (Robert, F.S.A.) The Electric Telegraph. With 200 Illustrations. 8vo, 12s. 6d.

SACRISTAN'S HOUSEHOLD (The). By the Author of "Mabel's Progress." Crown 8vo, 6s.

SAINT ABE AND HIS SEVEN WIVES. A Tale of Salt Lake City. Third and enlarged Edition. Crown 8vo, 5s.

SAINT PAUL'S MAGAZINE (The): Light and Choice. One Shilling Monthly. Half-yearly Vols., 10s. 6d. each.

SANDFORD AND MERTON. With Illustrations. 18mo, cloth gilt extra, 1s. 6d.

SAPHIR'S (Rev. Adolph) Conversion, Illustrated from Examples recorded in the Bible. Small 8vo, 3s. 6d.

SAUNDERS' (Katherine) The Haunted Crust, and other Stories. Two Vols., post 8vo, 21s.

SAVING KNOWLEDGE, Addressed to Young Men. By Thomas Guthrie, D.D., and W. G. Blaikie, D.D. Crown 8vo, 3s. 6d.

SAYER'S (Thos A.) Aids to Memory. A Practical System of Mnemonics. 12mo, 1s.

SEN'S (Baboo Keshub Chunder) Lectures and Tracts. Edited by S. D. Collet. Crown 8vo, 5s.

—————— English Visit. An authorized Collection of his principal Addresses delivered in England. Edited by S. D. Collet. Crown 8vo, 9s.

SHAEN'S (Mrs. William) School Lessons in Household Economy. Sewed, 6d.

SHELMERDINE'S (W.) Selection of the Psalms and other Portions of Scripture, arranged and marked for Chanting. Small 8vo, 1s.

—————————— One Hundred and Eighty Chants, Ancient and Modern. Selected from the most famous Composers, arranged for Four Voices, with Organ and Pianoforte Accompaniment. Crown 8vo, 2s. 6d.

SHORTREDE'S (Major-Gen.) Azimuth, Latitude, and Declination Tables. Demy 8vo, 7s. 6d.

SIMCOX'S (G. A.) Poems and Romances. Crown 8vo, 6s.

SMEDLEY'S (M. B.) Poems. Crown 8vo, 5s.

—————— Other Folk's Lives. Crown 8vo, cloth gilt extra, 5s.

—————— Linnet's Trial. Crown 8vo, cloth gilt extra, 5s.

SMEDLEY'S (Frank E.) Gathered Leaves. A Collection of the Poetical Writings of the late Frank E. Smedley. With a Memorial Preface, Portrait, &c. Imperial 16mo, cloth gilt, 8s. 6d.

SMITH'S (Alexander) Alfred Hagart's Household. Crown 8vo, 6s.

—————— A Summer in Skye. Crown 8vo, 6s.

—————— Dreamthorp. A Book of Essays written in the Country. Crown 8vo, 3s. 6d.

SMITH'S (David) Tales of Chivalry and Romance. With Illustrations. Crown 8vo, cloth gilt extra, 3s. 6d.

SMYTH'S (Professor C. Piazzi) Our Inheritance in the Great Pyramid. With Photographs and Illustrations. Square 8vo, 12s.

SMYTH'S (WARINGTON W., M.A., F.R.S.) Treatise on Coal and Coal Mining. With Illustrations. Post 8vo, 7s. 6d.

SPEN'S (KAY) True of Heart. Crown 8vo, 5s.

—— Tottie's Trial. Crown 8vo, 10s. 6d.

SPURGEON'S (REV. C. H.) The Saint and his Saviour; or, the Progress of the Soul in the Knowledge of Jesus. Crown 8vo, 3s. 6d.

STANLEY'S (DEAN) Scripture Portraits and other Miscellanies. Crown 8vo, 6s.

STAUNTON'S (HOWARD) The Great Schools of England: an Account of the Foundations, Endowments, and Discipline of the chief Seminaries of Learning in England. New Edition, with Account of all the Endowed Grammar Schools of England and Wales. Crown 8vo, 7s. 6d.

STEVENSON'S (REV. W. FLEMING) Praying and Working. Crown 8vo, 3s. 6d.; small 8vo, 2s.

STEWART'S (L.) The Wave and the Battle Field: Adventures by Sea and Land. With Illustrations. Crown 8vo, cloth gilt extra, 3s. 6d.

STIER'S (RUDOLF, D.D.) The Words of the Angels. Crown 8vo, 3s. 6d.

STUDIES IN FRENCH PROSE. Specimens of the Language from the Seventeenth Century to the Present Time. With Chronological and Critical Notices, Explanatory Notes, &c. 12mo, 3s. 6d.

STUDIES IN FRENCH POETRY. Specimens of the Language from the Seventeenth Century to the Present Time. With Chronological and Critical Notices, Explanatory Notes, &c. 12mo, 3s. 6d.

SUNDAY EVENING BOOK (THE). Short Papers for Family Reading. By James Hamilton, D.D., A. P. Stanley, D.D., John Eadie, D.D., Rev. W. M. Punshon, Rev. Thomas Binney, Rev. J. R. Macduff, D.D. Cloth antique, 1s. 6d.

SUNDAY MAGAZINE (THE). Edited by Thomas Guthrie, D.D., and W. G. Blaikie, D.D. 6d. monthly, Illustrated. Yearly Volumes, cloth gilt extra, 1865 to 1871, 8s. 6d. each; 1872, 7s. 6d.

SWAIN'S (CHARLES) Art and Fashion; with other Songs and Poems. Post 8vo, 7s. 6d.

TAINE'S (H., D.C.L.) Notes on England. Translated by W. F. Rae, with an Introduction by the Translator. Crown 8vo, 7s. 6d.

TAIT'S (GILBERT) The Hymns of Denmark. Rendered into English. Small 8vo, cloth gilt extra, 4s. 6d.

TANGLED TALK. An Essayist's Holiday. Post 8vo, 7s. 6d.

TATE'S (W.) Elements of Commercial Arithmetic. New Edition. 12mo, 2s. 6d. Key, 3s. 6d.

TAYLOR'S (BAYARD) Faust. A Tragedy. By Johann Wolfgang Von Goethe. Translated in the original metres. Two Vols., post 8vo, 28s.

TENNYSON'S (ALFRED) Poems. Small 8vo, 9s.

———————— Maud, and other Poems. Small 8vo, 5s.

———————— In Memoriam. Small 8vo, 6s.

———————— The Princess. Small 8vo, 5s.

———————— Idylls of the King. Small 8vo, 7s.

———————————————— Collected. Small 8vo, 12s.

———————— Enoch Arden, etc. Small 8vo, 6s.

———————— The Holy Grail, and other Poems. Small 8vo, 7s.

———————— Gareth and Lynette, &c. Small 8vo, 5s.

Library Edition of Mr. Tennyson's Works, Six Vols, post 8vo, 10s. 6d. each.

Pocket-volume Edition of Mr. Tennyson's Works. Eleven vols, 18mo, in neat case, 50s.; in extra binding, 55s.

Do. do. Eleven vols. 18mo., in blue and gold, sold as under :—

Poems. 3 vols., 15s.	Enoch Arden. 6s.
Maud, and other Poems, 5s.	In Memoriam. 6s.
The Princess. 5s.	Gareth, &c. 5s.
Idylls of the King. 3 vols. 15s.	

TENNYSON'S (ALFRED) Selections. Square 8vo, cloth extra, 5s.; gilt edges, 6s.

   Songs. Square 8vo, cloth extra, 5s.

   Concordance. Crown 8vo, 7s. 6d.

   The Window; or, the Songs of the Wrens. A Song-cycle by Alfred Tennyson, with Music by Arthur Sullivan. 4to, cloth gilt extra, 21s.

THOROLD'S (REV. A. W.) The Presence of Christ. Crown 8vo, 3s. 6d.

———————— On the Loss of Friends. Sewed, 3d.

———————— On Being Ill. Sewed, 2d.

THOUGHTS ON RECENT SCIENTIFIC CONCLUSIONS and their Relation to Religion. Crown 8vo.

THRONE OF GRACE (THE). By the Author of "The Pathway of Promise." Cloth antique, 2s. 6d.

TOUCHES OF NATURE. By Eminent Artists and Authors. Imperial 4to, cloth gilt extra, 21s.

TREASURY OF CHOICE QUOTATIONS. Crown 8vo, cloth extra, 3s. 6d.

TULLOCH'S (PRINCIPAL) Beginning Life. A Book for Young Men. Crown 8vo, cloth extra, 3s. 6d.

TYTLER'S (C. C. FRASER) Jasmine Leigh. Crown 8vo, cloth extra, 5s.

———————— Margaret. Two Vols., crown 8vo, 21s.

TYTLER'S (M. FRASER) Tales of Many Lands. With Illustrations. New Edition. Small 8vo, cloth gilt extra, 3s. 6d.

TYTLER'S (SARAH) The Songstresses of Scotland. Two Vols., post 8vo, 16s.

———————— Citoyenne Jacqueline. A Woman's Lot in the Great French Revolution. Crown 8vo, cloth gilt extra, 5s.

TYTLER'S (SARAH) Days of Yore. Crown 8vo, cloth gilt extra, 5s.

———— Girlhood and Womanhood. Crown 8vo, cloth gilt extra, 5s.

———— Papers for Thoughtful Girls. With Illustrations by Millais. Crown 8vo, cloth gilt extra, 5s.

———— Heroines in Obscurity. A Second Series of "Papers for Thoughtful Girls." Crown 8vo, cloth gilt extra, 5s.

———— The Diamond Rose. A Life of Love and Duty. Crown 8vo, cloth gilt extra, 5s.

———— The Huguenot Family in the English Village. With Illustrations. Crown 8vo, 6s.

———— "Noblesse Oblige." An English Story of To-day. Crown 8vo, 6s.

VAUGHAN'S (C. J., D.D.) Last Words in the Parish Church of Doncaster. Crown 8vo, 3s. 6d.

———— Earnest Words for Earnest Men. Small 8vo, 3s. 6d.

———— Characteristics of Christ's Teaching. Small 8vo, 2s. 6d.

———— Christ the Light of the World. Small 8vo, 2s. 6d.

———— Plain Words on Christian Living. Small 8vo, 2s. 6d.

———— Voices of the Prophets on Faith, Prayer, and Human Life. Small 8vo, 2s. 6d.

———— Half-hours in the Temple Church. Small 8vo, 3s. 6d.

———— Sundays in the Temple. Small 8vo, 3s. 6d.

———— Family Prayers. Crown 8vo, 3s. 6d.

———— The Presence of God in his Temple. Small 8vo, 3s. 6d.

VINET'S (ALEXANDER) Outlines of Philosophy. Edited by M. Astié. Post 8vo, 6s.

VINET'S (ALEXANDER) Outlines of Theology. Edited by M. Astié. Post 8vo, 6s.

WARING'S (A. L.) Hymns and Meditations. Cloth antique, 2s. 6d. Sewed, 1s.

WARREN'S (JOHN LEICESTER) Rehearsals: A Book of Verses. Crown 8vo, 6s.

——————— Philoctetes. A Metrical Drama after the Antique. Crown 8vo, 4s. 6d.

——————— Orestes. A Metrical Drama after the Antique. Crown 8vo, 4s. 6d.

WATSON'S (FORBES, M.R.C.S.) Flowers and Gardens. Notes on Plant Beauty. Crown 8vo, 5s.

WATTS' AND RIPPON'S HYMNS. Bound in One Volume, 32mo, roan embossed, 3s.; gilt edges, 3s. 6d.

WENTWORTH'S (PAUL) Amos Thorne and other Poems. Small 8vo, 3s.

WHEELER'S (J. TALBOYS, F.R.G.S.) Historical Geography of the Old and New Testaments. Folio, 7s. 6d.

——————— Analysis and Summary of Old Testament History and the Laws of Moses. Post 8vo, 5s. 6d.

——————— Analysis and Summary of New Testament History. Post 8vo, 5s. 6d.

——————— Popular Abridgment of Old and New Testament History. Two Vols., 18mo, 2s. each.

WHITEHEAD'S (REV. H.) Sermons, chiefly on Subjects from the Sunday Lessons. Crown 8vo, 6s.

WHITNEY'S (ADELINE T.) Pansies. "——— for Thoughts." Square 8vo, 2s. 6d.

WILBERFORCE'S (BISHOP) Heroes of Hebrew History. New and Cheaper Edition. Crown 8vo, 5s.

WILDE'S (ROBERT) Poems. Edited by the Rev. John Hunt. Small 8vo, 3s. 6d.

WILKINSON'S (Rev. W. F.) Personal **Names** in the Bible. Small 8vo, 6s.

WILLEMENT'S (E. E.) Familiar **Things**: their History, &c. Small **8vo, 2s. 6d.**

WILLIAMS' (Sarah) Twilight Hours. A **Legacy** of Verse. With a Memoir by E. H. Plumptre, M.A. Third and enlarged Edition. Crown 8vo, 5s.

WINDWAFTED SEED. Edited by Norman Macleod, D.D., **and Thomas** Guthrie, D.D. Crown 8vo, **3s. 6d.**

WORBOISE'S (E. J.) Sir Julian's Wife. Small 8vo, 5s.

———— The Wife's Trials. Small 8vo, 3s. 6d.

———— The Life of **Thomas Arnold, D.D.** Small 8vo, 3s. **6d.**

———— Campion Court. A Tale of the Days of the Ejectment Two Hundred **Years Ago.** Crown 8vo, 5s.

———— The Lillingstones of Lillingstone. Crown 8vo, 5s.

———— **Lottie** Londsdale; or the Chain and its Links. Crown 8vo, 5s.

———— Evelyn's **Story**; or Labour and Wait. Crown 8vo, 5s.

WORDSWORTH'S Poems for the Young. With Illustrations. Square 8vo, cloth gilt extra, **3s. 6d.**

YORKE'S (Onslow) The **Story of** the International. Crown 8vo, 2s.

YOUNG'S (John, LL.D.) The Christ of History. New and enlarged Edition. Crown 8vo, 6s.

———— The Life and Light of Men. Post 8vo, 7s. 6d.

———— The Creator and the Creation, how related. Crown 8vo, 6s.

www.ingramcontent.com/pod-product-compliance
Lightning Source LLC
Chambersburg PA
CBHW031348230426
43670CB00006B/469